AMERICAN MILITARY
TECHNOLOGY

AMERICAN MILITARY TECHNOLOGY

THE LIFE STORY OF A TECHNOLOGY

Barton C. Hacker, with the assistance of Margaret Vining

GREENWOOD TECHNOGRAPHIES

GREENWOOD PRESS
Westport, Connecticut • London

Library of Congress Cataloging-in-Publication Data

Hacker, Barton C., 1935–
  American military technology : the life story of a technology / Barton C. Hacker,
with the assistance of Margaret Vining.
      p.   cm. — (Greenwood technographies, ISSN 1549-7321)
  Includes bibliographical references and index.
  ISBN 0–313–33308–4 (alk. paper)
    1. Military art and science—United States—Technological innovations.
2. Technology—United States. I. Vining, Margaret. II. Title. III. Series.
  U43.U4H33   2006
  623.0973—dc22   2005033544

British Library Cataloguing in Publication Data is available.

Library of Congress Catalog Card Number: 2005033544
ISBN: 0–313–33308–4
ISSN: 1549–7321

First published in 2006

Greenwood Press, 88 Post Road West, Westport, CT 06881
An imprint of Greenwood Publishing Group, Inc.
www.greenwood.com

Printed in the United States of America

The paper used in this book complies with the
Permanent Paper Standard issued by the National
Information Standards Organization (Z39.48–1984).

10 9 8 7 6 5 4 3 2 1

Contents

Series Foreword

In today's world, technology plays an integral role in the daily life of people of all ages. It affects where we live, how we work, how we interact with each other, and what we aspire to accomplish. To help students and the general public better understand how technology and society interact, Greenwood has developed *Greenwood Technographies*, a series of short, accessible books that trace the histories of these technologies while documenting *how* these technologies have become so vital to our lives.

Each volume of the *Greenwood Technographies* series tells the biography or "life story" of a particularly important technology. Each life story traces the technology, from its "ancestors" (or antecedent technologies), through its early years (either its invention or development) and rise in prominence, to its final decline, obsolescence, or ubiquity. Just as a good biography combines an analysis of an individual's personal life with a description of the subject's impact on the broader world, each volume in the *Greenwood Technographies* series combines a discussion of technical developments with a description of the technology's effect on the broader fabric of society and culture—and vice versa. The technologies covered in the series run the gamut from those that have been around for centuries—firearms and the printed book, for example—to recent inventions that have rapidly taken over the modern world, such as electronics and the computer.

While the emphasis is on a factual discussion of the development of the technology, these books are also fun to read. The history of technology is full of fascinating tales that both entertain and illuminate. The authors—all experts in their fields—make the life story of technology come alive, while also providing readers with a profound understanding of the relationship of science, technology, and society.

Introduction

From their earliest days, American military, engineering, and scientific institutions have interacted with each other in manifold ways, with consequences reaching far beyond the institutions themselves. Since World War II, military managers have gained control of unprecedented resources and have learned to harness science and engineering to their wants. Imperfect as yet and enormously expensive, these new techniques have nonetheless produced a striking panoply and forced great changes throughout the armed forces. They have also transformed American science and engineering, reshaped American universities, and strongly affected almost every aspect of American society.

Technological innovation of almost every type has historically answered more to military purpose than is commonly allowed. This is not simply a matter of technological change fostered by wartime demands. Although war has often served as a forcing bed for military-technological innovation, war cannot be the centerpiece of a history of military technology. The sporadic impact of a specific war or wars, or even of war in general—the fostering of technological change by wartime demands, for instance—pales in comparison with the powerful and persistent, if not always obvious, interaction of military with other social institutions over extended periods of time. And influence always flowed both ways: society's impact on technology is no less important than technology's on

society. Changing technologies, whether explicitly military, military sponsored, or military influenced, have exerted profound institutional effects.

Political revolution in America occurred as industrial revolution was
beginning to transform England and shape the modern world. The United
States was very much a product of the industrial era, as machine production
came increasingly to dominate war planning and conduct. These trends
were just emerging at the time of the Civil War, which displayed much
that was old as well as much that was new.

During the late nineteenth and early twentieth centuries, military-
technological change far more rapid than ever prevailed under preindustrial
conditions became the rule. More telling, the rate of change increased. As
time between idea and product grew ever shorter, the pace of meaningful military-technological change accelerated. The United States, which
had been a leader in military-technological innovation until 1865, fell
behind in ensuing decades, though individual Americans were significant
contributors.

Despite improvements in military and naval technology so great as to
suggest another military revolution, the most striking feature of the late
nineteenth and early twentieth centuries was an extraordinary expansion of
productive capacity. Industrialized states could produce so much in so little
time with so few workers that armed forces no longer seemed able to
enforce decision. When armies upon armies could be raised, equipped, and
maintained, war became nothing so much as a giant siege operation, as the
First World War showed to devastating effect.

The twenty-year Armistice between the world wars proved to be a
time of intense ferment in military affairs. Strategic aerial warfare, armored
warfare, amphibious warfare, carrier aviation, all became subjects of intense
discussion and experimentation, organizationally as well as technologically.
In World War II these and further technical and tactical innovations bore
fruit, yielding opportunities for generalship comparable to the Napoleonic
wars. Mechanized armed forces wedded to the tactics of infiltration
restored maneuver to battle and decisiveness to war. Striking differences in
the course and outcome of the two world wars reflected the success of
interwar efforts to harness scientific research to military needs.

No less important than the practical results from 1939 onward were the
new forms for mobilizing science, in which the United States once again
assumed a leading role. Arrangements made to exploit research in World
War II permanently transformed relations among American military,
technological, and scientific institutions. Academic and industrial laboratories largely dependent on government funding became major sources of
military-technological innovation during the later twentieth century. The

reluctant military innovators of the later nineteenth century became enthusiastic seekers of technological novelty in the later twentieth.

The constant military problem has always been how to prevail in the field against comparably equipped armies. During the modern period, perhaps especially in the United States, one way out of the impasse often seemed to be technological innovation—that is, striking "comparably equipped" from the equation. This was the dream of decisive secret weapons that the Manhattan Project and development of the atomic bomb seemed to confirm. Directed research and development promised a constant edge to the nation able to sustain so costly an effort. The United States adopted what was, in effect, a policy of permanent technological revolution.

The institutionalization of these new relationships dominated the immediate postwar period. For two decades after the war, nuclear weapons and their delivery systems remained the technological focus. Working with the armed forces, the newly established Atomic Energy Commission steadily improved fission weapons and developed thermonuclear weapons. Each leg of the strategic triad that was in place by 1960 and steadily improved thereafter—long-range manned bombers, intercontinental ballistic missiles, and submarine-launched intermediate-range missiles—could deliver a devastating nuclear attack. The first early warning and reconnaissance satellites reached orbit in the 1960s and their capabilities expanded during succeeding decades. The even greater effort that went into finding defenses against missiles achieved noticeably less success.

War in Vietnam did much to shift the military-technological focus back to conventional weapons, a term which had expanded to include a wide range of electronic sensor and guidance technologies. But the overwhelming superiority of American military technology in Vietnam did not translate into American victory, a failure that caused many younger officers and other analysts to rethink the relations among technology, tactics, doctrine, and policy. Although military-technological innovation continued apace, its products acquired a new context displayed with startling clarity in the Persian Gulf and the Balkans as the twentieth century ended.

Attention to the changing social and political contexts for military-technological change has become a major feature of the modern history of military technology. Understanding the technical parameters, however important, can never be enough. Just as military institutions are but varieties of social institutions, so too military-technological change is but a variety of social change.

This marks an important shift from the traditional history of military technology, which favored a nuts-and-bolts approach focused on the thing itself—weapon, machine, fortification, all the physical relics of war-making. The history of military technology now displays a marked concern for the social context of its subject and a penchant for asking questions that extend well beyond the narrow bounds of describing battles or hardware. The expanding concept of military technology has now come to embrace ideas and institutions, which, like tools and machines, must be fabricated. Studying organization, management, and doctrine has become as much a part of the field as studying weapon development.

Despite many good specialist studies, broad attempts to relate weaponry to social context have not been particularly successful, and no one has tried to write a contextual history of American military technology. This book is offered as a preliminary attempt to repair that lack, placing the history of American military technology and science in the context of American social, political, and economic development.

Timeline

1813	Simeon North awarded contract for 20,000 pistols with interchangeable parts
1814	Robert Fulton's *Demologos* (or *Fulton I*) launched; first steam-powered warship
1817	Sylvanus Thayer appointed superintendent of West Point
1818	Army Medical Corps established
1821	George Bomford becomes chief of ordnance
1824	U.S. Army Corps of Engineers authorized to undertake civil as well as military construction projects
1830	Baltimore & Ohio Railroad in operation
1830	U.S. Navy Depot of Charts and Instruments established
1838	Corps of Topographical Engineers established
1838–1842	United States Exploring Expedition
1841	U.S. Army's first ordnance manual issued
1842	Springfield Armory's Model 1842 smoothbore caplock musket, first product of the uniformity system from a government arsenal
1842	Matthew Fontaine Maury appointed superintendent of the depot of charts and instruments in the Navy Department
1843	Alexander Dallas Bache appointed superintendent of the United States Coast Survey
1843	USS *Princeton* launched, world's first warship driven by stern-mounted screw propeller
1845	U.S. Naval Academy founded at Annapolis, MD
1846–1848	American War with Mexico
1851	Great Exhibition in London
1852–1854	Crimean War
1855	U.S. Model 1855 rifle-musket, first standard issue rifled firearm
1857	Introduction of 12-pounder field gun, the "Napoleon"
1860	U.S. Army Signal Corps, world's first specialized tactical military communications organization
1860	Spencer repeating rifle patented
1861–1865	American Civil War

1861	Civilian U.S. Sanitary Commission established to aid Army Medical Department
1861	Employment of women as paid nurses in army hospitals authorized
1861	Balloon Corps of the Army of the Potomac established
1862	CSS *Virginia* (or *Merrimac*) and USS *Monitor*, first battle between armored steam-powered warships
1862	Military training in colleges authorized by Morrill Act
1862	Richard Gatling patents his multiple-barrel machine gun
1863	Union victories at Vicksburg and Gettysburg
1863	Establishment of National Academy of Sciences
1864	First successful submarine attack, CSS *Hunley* using a spar torpedo
1864	U.S. Army Ambulance Corps established
1864	Sherman's March to the Sea and Sheridan's devastation of the Shenandoah Valley
1866	U.S. Army adopts Gatling gun
1870–1871	Franco-Prussian War and German unification
1870–1890	Weather Service managed by Signal Corps
1881	American Red Cross established
1881	School of Application for Cavalry and Infantry, Fort Leavenworth, KS, established
1883	Congress approves building the navy's first steel ships
1884	Naval War College, Providence, RI
1884	Hiram Maxim patents machine gun
1885	Publication of Henry Metcalfe, *The Cost of Manufactures and the Administration of Workshops, Public and Private*
1886	French chemist Paul M. E. Vielle invents smokeless powder
1886	French Lebel, first smallbore rifle using smokeless powder
1886–1918	U.S. Army management of National Park System
1887	U.S. Army Hospital Corps established
1890	Publication of Alfred Thayer Mahan, *The Influence of Sea Power upon History*

1890	U.S. Navy launches its first modern warship, an armored cruiser
1892	U.S. Army adopts Danish Krag-Jorgensen .30-caliber magazine rifle firing smokeless cartridges
1893	Army Medical School established
1893	George Squier receives Ph.D. in physics from Johns Hopkins University
1897	French Model 1897 75-mm gun introduced
1898–1902	Spanish-American War and Philippine War
1899	David Taylor named director of the Experimental Model Basin
1899–1902	Boer War
1900	Army War College established
1900	U.S. Navy purchases *Holland I*, world's first military modern submarine
1901	Walter Reed's Yellow Fever Commission in Cuba proves the disease's cause to be a mosquito-borne virus
1901	Army Nurse Corps established
1902	William Sims appointed Inspector of Target Practice
1903	First U.S. semiautomatic rifle introduced, Springfield .30-caliber
1903	U.S. Army Model 1902 3-inch field gun introduced
1903	Wright Brothers' first powered flight, Kitty Hawk, NC
1904	William Crawford Gorgas appointed chief sanitary officer of the Panama Canal Zone
1905	U.S. Army's first signals school at Fort Leavenworth, KS
1906	HMS *Dreadnought* launched
1908	Signal Corps contracts for Wright Flyer, first military aircraft purchase
1908	Navy Nurse Corps established
1909	USS *Delaware*, America's first dreadnought, launched
1911	Vaccination for typhoid made compulsory for U.S. Army
1914	Opening of Panama Canal
1914–1918	First World War

1915   Naval Consulting Board established

1916   National Advisory Committee on Aeronautics (NACA) established

1916   Reserve Officer Training Corps (ROTC) established

1916   Council of National Defense established

1916   National Academy of Sciences creates National Research Council
              (NRC)

1916–1917   Mexican punitive expedition

1917–1918   U.S. declaration of war

1917   George Squier appointed Chief Signal Officer

1917   William Sims named head of U.S. naval forces in Europe

1917   Samuel Rockenbach named to head independent U.S. Army Tank
              Corps

1917   Bolshevik revolution in Russia

1918   U.S. Army Tank Corps established

1918   U.S. Army Chemical Warfare Service established

1918–1919   Great flu epidemic

1919   Julia Stimson named dean of the Army School of Nursing and su-
              perintendent of Army Nurse Corps

1920   Congress abolishes the independent Tank Corps and Chemical
              Warfare Service

1921   Battleship sinking by aerial bombs arranged by Billy Mitchell

1921   U.S. Navy Bureau of Aeronautics established, headed by William
              Moffett

1921   Washington Naval Treaty

1922   USS *Langley* commissioned, U.S. Navy's first experimental aircraft
              carrier

1922   Curtiss NAF TS-1 entered service, first U.S. airplane specifically
              designed for carrier-based operation

1923   Naval Research Laboratory established

1924   George W. Lewis named director of aeronautical research

1926   First liquid-fuel rocket launched by Robert H. Goddard

1927	USS *Lexington* and USS *Saratoga* commissioned; first U.S. fleet aircraft carriers
1928	U.S. Army experimental mechanized force formed
1931	Norden bombsight introduced
1935	U.S. Marine Corps, *Tentative Manual of Landing Operations*, issued
1938	U.S. Navy procures first LCVP (Landing Craft Vehicle and Personnel), a modified version of Andrew Higgins' "Eureka" boat
1938	Prototype radar system installed on battleship USS *New York*
1939	First Boeing B-17 Flying Fortress delivered
1939–1945	World War II
1940	Battle of Britain
1940	Sikorsky VS-300 helicopter completes first successful free flight by a single-rotor helicopter
1940	National Defense Research Committee (NDRC) established, headed by Vannevar Bush
1940	U.S. Marine Corps procures the first LVT (Landing Vehicle Tracked), a modified version of Donald Roebling's "Alligator"
1940	U.S. Army Armored Force established, headed by Adna Romanza Chaffee, Jr.
1941	Office of Scientific Research and Development (OSRD) established, headed by Vannevar Bush
1941	MIT Radiation Laboratory established
1941	American entry into World War II
1942	Pentagon opens for business just sixteen months after construction began
1942	Applied Physics Laboratory, Johns Hopkins University, established, headed by Merle Tuve
1942	Manhattan Engineer District established, headed by Leslie H. Groves
1942	First proximity-fuzed shells distributed
1942	Bazooka standardized and distributed to troops
1942	First sustained chain reaction in uranium, University of Chicago
1942	First *Essex*-class carriers commissioned

1942	Publication of Alexander de Seversky, *Victory through Air Power*
1943	Nuclear weapons laboratory established at Los Alamos, NM
1943	Naval Ordnance Test Station (NOTS), China Lake, CA, established
1944	Jet Propulsion Laboratory established
1944	HVAR distributed to combat units
1944	ME-262, first operational jet fighter
1945	Atomic bombing of Hiroshima and Nagasaki
1945	Vannevar Bush report to the president, *Science—The Endless Frontier*
1946	Operation Crossroads, Bikini Atoll, A-bomb demonstration
1946	Lockheed F-80 Shooting Star, first operational U.S. jet fighter
1946	Office of Naval Research established
1946	Air corps contracts for Project RAND with Douglas Aircraft Company, Santa Monica, CA, later the Rand Corporation
1946	Strategic Air Command (SAC) established
1946	Sandia Laboratory, Albuquerque, NM, established
1946	Publication of Ruth Benedict, *The Chrysanthemum and the Sword*
1947	Department of Defense (DOD) established, uniting the armed forces in one structure
1947	Department of the Air Force separated from Department of the Army
1947	Atomic Energy Commission (AEC) established
1947	Armed Forces Special Weapons Project (AFSWP) established
1948	Weapons Systems Evaluation Group (WSEG) established in Pentagon
1948	National Institutes of Health established
1949	First Soviet A-bomb tested
1949	Publication of Vannevar Bush, *Modern Arms and Free Men: A Discussion of the Role of Science in Preserving Democracy*
1950	Crash H-bomb development program
1950	National Science Foundation established

1950	National Security Directive NSC-68 issued, Cold War policy statement
1950–1953	Korean War
1951	Aircraft Nuclear Propulsion program initiated
1951	Army Research Office established
1951	Nevada Test Site for nuclear weapons becomes operational
1952	Lawrence Livermore Laboratory established
1952	DEW (Distant Early Warning) Line initiated
1952	Air Force Office of Scientific Research established
1952	Publication of Bernard Brodie, *The Atom Bomb*
1954	USS *Nautilus* launched, first nuclear-powered submarine
1954	Operation Castle, Enewetak Atoll, proof-tests of practical H-bomb
1955	Boeing B-52 Stratofortress long-range jet bomber becomes operational
1955	Navy Sidewinder and air force Sparrow air-to-air missiles become operational
1955	U.S. Army choice of Bell Model 204 as its utility helicopter
1955	U.S. Navy Special Projects Office established
1956	Institute for Defense Analyses succeeds WSEG
1956	Responsibility for all long-range missile programs assigned to air force
1957	Soviet Sputnik launched; first artificial satellite
1958	Activation of the first operational Atlas ICBM squadron, Vandenberg Air Force Base, CA
1958	Nike Hercules declared operational
1958	National Aeronautics and Space Administration (NASA) established
1958	President's Science Advisory Committee established; James Killian named presidential special assistant for science and technology
1958	Directorate of Defense Research and Engineering established
1958	Advanced Research Projects Agency (ARPA) established, later renamed Defense Advanced Research Projects Agency (DARPA)
1958	National Defense Education Act (NDEA) becomes operational

1959	First Corona satellite launched
1959	First deliveries of UH-1A (Huey) turbine-powered helicopters
1959	Safeguard antiballistic missile system initiated
1960	First version of COBOL programming language completed, under leadership of Grace Hopper
1960	First Polaris submarine, USS *George Washington*, on patrol
1960	First SAMOS (Satellite and Missile Observation System) launched
1960	First meteorological satellite launched, Tiros I (Television InfraRed Observation Satellite)
1960	First navigational satellite orbited, Transit I-B
1960	Ballistic Missile Early Warning System (BMEWS) operational, a network of ground stations to warn of missile attack
1960	National Reconnaissance Office established
1960	John Boyd manual, "Aerial Attack Study"
1960	Publication of Charles J. Hitch and Ronald J. McKean, *The Economics of Defense in the Nuclear Age*
1961	President Eisenhower's farewell address warns of military-industrial complex
1961	Robert S. McNamara named Secretary of Defense
1961	Charles J. Hitch named Assistant Secretary of Defense and Defense Comptroller
1962	First active communication satellite launched, Telstar I
1962	Defense Meteorological Satellite Program initiated
1963	First two Vela nuclear radiation detection satellites launched
1963	SAGE declared operational
1964	Lockheed SR-71 Blackbird becomes operational
1964	Satellite navigation system becomes operational, Transit
1965–1973	Vietnam War
1966	Defense Communications Planning Group (DCPG) established
1966	Project Hindsight report issued
1967	First combat use of Walleye electro-optically guided bomb

1968	IDCSP (Initial Defense Communications Satellite Program) declared operational
1968	Project TRACES (Technology in Retrospect and Critical Events in Science) report issued
1969	Passage of the Mansfield Amendment
1971	First KH-9 Big Bird high-resolution photoreconnaissance satellite launched
1972	First NASA Earth Resources Technology satellite launched, later renamed LandSat
1972	Air attack on Thanh Hoa bridge, first combat use of laser-guided bombs
1972	ABM Treaty
1973	First NAVSTAR (Navigation Satellite Timing and Ranging) satellite launched
1973	DSP (Defense Support Program) early warning satellite system declared operational
1975	Safeguard system declared operational and dismantled
1975	Defense Support Program (DSP) operational
1976	KH-11 (Key Hole) photoreconnaissance satellite launched
1979	First F-16 Fighting Falcons accepted for air force service
1982	First satellite in Defense Satellite Communications System launched
1983	First Patriot antimissile battalion operational
1983	Lockheed F-117A enters service
1983	Strategic Defense Initiative (SDI) initiated
1984	First Tomahawk precision-guided land attack cruise missiles deployed
1988	Global positioning systems operational, NAVSTAR
1991	Dissolution of Soviet Union
1991	Gulf War I
1991	First combat use of Tomahawk cruise missiles
1994	First Milstar (Military Strategic/Tactical and Relay System) advanced military communication satellite launched

1994   Joint Defense, Commerce, and NASA weather satellite program established

1998   Joint Department of Defense/NASA office for LandSat-7 established

1999   Air war in Kosovo

1

Before Industrialization: Through the Early Nineteenth Century

◆

Preindustrial European military technology transplanted to America exhibited certain persistent traits throughout the period from the Spanish conquest through the early nineteenth century. Soldiering during this era is perhaps best understood as a preindustrial craft, jealously guarding the tools of its trade and slow to accept innovation. Although still glacial by later standards, Western military technological change had nonetheless already begun to outpace its nonwestern contemporaries. Yet the Western edge, with a few notable exceptions, remained modest before the nineteenth century.

If soldiering was a craft in early modern Europe, weapons were the tools of the trade. They required individual expertise to wield effectively and individual experience to improve. Practical innovations in weaponry were mainly incremental, coming from the actual users as filtered through other craftsmen, for weapon-making, too, rested on craft traditions. Like other preindustrial technologies, it improved by rule-of-thumb, cut-and-try methods. Nothing resembling organized research existed, though individual practitioners doubtless sometimes experimented with new processes or designs.

Accordingly, the pace of technological change in weaponry, as in all other preindustrial technologies, was slow. Putting it another way, troops carried much the same weapons from one decade to the next, or even one century to the next. That left plenty of time for tacticians to devise, via the same cut-and-try approaches that marked weapon development, appropriate means of using arms and combining them effectively. That was beginning to change in the early decades of the nineteenth century, and American military officers and civilian inventors played an unexpectedly large role. Military engineers led the construction of the new nation's infrastructure and the exploration of its western lands, while ordnance officers and civilian gunsmiths created an American system of interchangeable parts manufacturing that marked a major step on the way to mass production.

CONQUEST AND SETTLEMENT

Within particular military culture areas, such as western Europe or the Eastern Woodlands of North America, military technology was likely to achieve and maintain rough parity. Across culture areas, however, technological advantages might be decisive, as shown by European horses, steel, and guns in the New World. The uncontested supremacy of European ships and guns was the foundation of European conquest in America, though other aspects of Western military technology also contributed significantly. At sea the gun-carrying sailing ship remained the master military-technological system of the early modern era, right up to the mid-nineteenth century. It also remained a Western monopoly throughout the period.

Until late in the seventeenth century, conflicts between European forces in America were few and far between. Military interaction chiefly pitted Europeans, often with Native American allies, against other Native Americans. Colonial forces regularly exploited disunity among the Native Americans, who rarely saw themselves as a single people. Victory ordinarily crowned the most advanced arms, yet that told only part of the story. The sustained ferocity of Euroamerican warfare against perceived foes multiplied the effects of their weapons. Over the course of time, Native Americans learned to modify their own approaches to the conduct of war. The technological gap also tended to narrow as Native Americans adapted European military technologies to their own purposes, especially when matchlock muskets gave way to cheaper, lighter, and more efficient flintlocks toward the end of the century.

Ultimately, European settlers came to enjoy other, perhaps more decisive, advantages over their indigenous rivals, most notably manpower. Eurasian diseases devastated virgin New World populations in New

England and Virginia, leaving large swaths of vacant land to be occupied by the first settlers. Disease also afflicted the invaders, but as products of the historically far more complex Eurasian disease environment, in population terms they suffered less and recovered more quickly. The balance of power increasingly favored the newcomers, whose agricultural economies could sustain larger numbers of people in a given area than could the hunting and gathering or horticultural economies that prevailed in Native Eastern North America. Euroamerican populations also swelled their numbers with ongoing emigration from the Old World.

Enlightenment military institutions underwent a sea change in the passage to America. Although the formal evolutions of British and French forces on the Plains of Abraham testified to the persistence of well-tried forms, close order drill and geometric precision could seldom be effectively maintained in the wilderness. Such circumstances produced a looser, more independent style that profoundly affected the conduct of war in America. As the classic sieges of Louisbourg by British and colonial forces in 1745 and again in 1758 amply demonstrated, military engineering found a home in the New World as well. Nor did the attack and defense of fortifications only characterize warfare among Europeans and colonists. Palisaded villages were also a feature of Native American life. British and colonial military engineers surveyed and built roads throughout the colonies, beginning the tradition that characterized the creation of the new nation's transportation network after independence.

For all that, fortification never attained the central importance in America that it did in Europe. Fortified cities almost everywhere in eighteenth-century Europe had become central to strategic planning, military engineering a crucial aspect of military organization. Siegecraft and field maneuvers alike took on a geometrical precision that eighteenth-century thinkers were wont to perceive as scientific. They talked less of "the art of war," more of "military art and science" or simply "military science," a usage that became normal in the nineteenth century. When military schools began to appear in the mid-eighteenth century, they stressed scientific-technical training, and in fact everywhere became the model for nonmilitary schools of science and engineering.

International rivalries reflected in the New World dominated events during the eighteenth century. American conflicts increasingly resembled those in Europe—that is, fought between forces more or less evenly matched in technological terms—despite some novelties stemming from the mutual interaction of Native Americans and Euroamericans. Although Euroamericans continued to enjoy some degree of technological advantage, their decisive edge increasingly became weight of numbers. Native

American populations suffered wave after wave of disease—as likely to precede as follow the westward movement of settlers, always debilitating and too often fatal—while the population of European origin grew steadily.

THE AMERICAN REVOLUTION

Gun making along traditional lines flourished in the colonies, and colonial naval stores became a major resource of British sea power. Interactions of military and naval engineering with other preindustrial American institutions, even as limited as they were, had virtually no counterpart in science and medicine. Colonial science foreshadowed the observational preoccupations that became so marked a feature of nineteenth-century American science. It also remained distinctly provincial, if not backward, and had no discernible effect on military institutions.

Much the same was true of medicine, still in the eighteenth century much more art and craft than science. Military surgeons and nurses might accompany larger expeditions, but hardly affected operations. Like all early modern armies, both British and American forces in the Revolutionary War relied on female camp followers as nurses. Medical services, too, remained little changed from earlier practices. Although the Continental Army acquired a medical department, and Washington's army undertook large-scale inoculations against smallpox, the long-standing pattern of ten deaths from disease for every death in battle showed little sign of abating.

America for the most part fought England with the standard weapons of the age. The basic firearm in America, as throughout the Western world in the late eighteenth century, was the flintlock musket, the perfected product of well over a century's evolution. Although individual weapons might vary in length, weight, and bore, each was a single-shot smooth-bore muzzleloader firing a heavy lead ball—fifteen to the pound was one standard recipe—about two-thirds inch in diameter. Individual paper cartridges held ball and gunpowder, the soldier tearing them open to load. This made for speedy firing. Trained troops could get off three or four shots a minute, or two volleys against a normal charge. To receive the charge, the soldier fitted a socket bayonet over its muzzle and converted the musket into a surrogate pike; this was the key late-seventeenth-century innovation that rendered the true pike superfluous.

Rifles hold a special place in Revolutionary folklore. Well known since the fifteenth century as a way to improve range and accuracy, rifling suffered a major military drawback. It demanded a tight-fitting bullet, which the soldier must ram down the barrel of his muzzleloader, a task far harder and slower than loading smoothbore muzzleloaders, especially after firing had fouled the rifling with the residue of black powder. Yet in the right circumstances, such as those that prevailed along the western colonial frontier in the early eighteenth century, the rifle's advantages might outweigh its shortcomings. Pennsylvania gunsmiths transformed the stubby, large-bore jäger rifle they knew from central Europe into the first distinctively American contribution to firearms technology, the legendary Pennsylvania-Kentucky rifle with its long barrel and relatively small bore. Despite the glamour that attached to the weapon then and now, it was not particularly effective in battle. For military purposes, the rifle's drawbacks restricted it to an essentially auxiliary role until well into the nineteenth century.

British control of the seas dictated the shape of the American war of independence. American ships could only harass, not challenge the enemy fleet. Although without impact on the course of the war, one novelty allowed the fledgling United States to stake an early claim on the technology of undersea warfare. David Bushnell of Connecticut designed and built a submersible one-man boat to attack British warships from below. Initially using his own funds, then partly subsidized by the Connecticut legislature— a pattern that would be repeated more than once in later American history— he completed building his boat in 1775. A year later he had his first target, the 64-gun *Eagle* moored in New York harbor. On the night of September 6, 1776, Sgt. Ezra Lee, whom Bushnell had personally trained, manned *Turtle* on her first sortie. He got as far as *Eagle*'s hull, but when his auger failed to pierce her bottom he could not attach the mine. Two later attempts achieved even less. Although *Turtle* undertook no more missions, she was not forgotten, and the dream of undersea warfare stayed alive.

David Bushnell (1742–1824)

A native of Connecticut, Bushnell graduated from Yale College in 1775. He immediately began work on a submersible one-man boat. Creatively adapting European ideas, he intended his muscle-powered *Turtle* to approach an enemy vessel unseen and attach a mine to her hull beneath the water line. Although it failed to accomplish its mission, the *Turtle* may fairly be said to mark the birth of the submarine.

1875 drawing of the *American Turtle* by Lt. Francis Barber. (U.S. Navy Photo)

Bushnell subsequently served as an officer in the Continental Army, in the company of sappers and miners that George Washington formed in 1779. His final assignment, in 1783, was commanding the engineer detachment at West Point, New York. After a lengthy sojourn in France, Bushnell returned to the United States to become a teacher and physician in Georgia, where he died.

Such attempts to attack the British command of the sea reflected the importance of logistics in deciding the war's outcome. So long as the British fleet controlled the sea-lanes, the American rebels had to rely almost entirely on their own resources. That almost proved their undoing. The largely agricultural colonies enjoyed ample food resources and succeeded reasonably well in building what military industry they needed almost from scratch. Distributing supplies, not finding them, was the crux of the matter and a constant problem for American forces. Ultimately, however, logistics proved the Achilles' heel of the British army in America. French intervention achieved decisive results by temporarily isolating British land forces from their support.

THE SIGNIFICANCE OF WEST POINT

Transportation and fortification remained constant problems for George Washington's armies, aggravated by a persistent lack of engineers. Scarcely had the United States achieved independence when Washington urged establishment of a military academy, citing the want of engineers and artillerists that had much plagued his army during the war. Although other motives played a part, the 1802 founding of the United States Military Academy at West Point owed much to such concerns. The same legislation that created West Point also founded the Corps of Engineers, which would direct the new academy.

An uncertain mission, internal conflicts, and inadequate staff plagued the military academy's first fifteen years. That all began to change with the appointment of Sylvanus Thayer as superintendent in 1817. A graduate of Dartmouth and West Point, Thayer had just concluded a two-year inspection tour of European military schools and installations. Following the example of the famous French engineering and artillery schools that the army had sent him to study, Thayer initiated the educational and administrative reforms that made West Point, during his sixteen-year tenure, America's preeminent school of engineering. It combined officer training with a highly technical undergraduate education.

Important as such skills might be for the army, they mattered even more for a young nation so notably lacking in them. West Point long remained America's only engineering school, and it more than held its own against competition from other schools before the Civil War. It always trained military engineers, but the same course of study also helped meet America's expanding demand for civil engineering. In fact, Congress soon designated West Point the National Engineering School. Graduates of the military academy, whether or not they remained in uniform, fanned out across the countryside to construct and maintain America's burgeoning transportation network. They surveyed roads and built forts, constructed canals and bridged streams, dredged harbors and cleared rivers. Sometimes they did the work themselves, often they supervised others; always they trained new hands, most of them not military. American civil engineering owed a great deal to West Point and military engineering.

In the early nineteenth century, such on-the-job training was the way most engineers learned their trade, but West Pointers also moved into formal education as teachers of science and engineering, or even as founders of engineering schools. West Point became the nation's major source not only of civil engineers, but also of engineering educators. Thayer himself founded the school of engineering at Dartmouth College, and the engineering textbooks of longtime West Point faculty member Dennis Hart Mahan became American standards of engineering instruction. In the three decades before the Civil War, West Pointers as teachers, writers, and practitioners fostered science and engineering at Dartmouth, Harvard, Yale, and other colleges. They also helped staff the new United States Naval Academy when it opened at Annapolis in 1845. The naval academy did much the same for mechanical engineering that West Point had done for civil engineering.

Engineers trained at West Point took leading roles in the early years of American railroad building. Two of the four-member engineering team hired by America's first major railroad, the Baltimore & Ohio, and nine of the ten topographical assistants were West Pointers. Most of these assistants became railroad engineers, forming the core of this new profession. Two West Pointers who became fast friends at the academy, William Gibbs McNeill and George Washington Whistler, cut their railroad engineering teeth on the Baltimore & Ohio as young army officers, then went on to other projects in the United States and abroad.

William Gibbs McNeill (1801–1853) and George Washington Whistler (1800–1849)

McNeill and Whistler began their lifelong friendship as cadets at West Point. After his first wife, Mary Smith, died very young, Whistler married McNeill's sister, Anna Matilda. Their son, the noted painter James McNeill Whistler, attended West Point without graduating.

Both McNeill and Whistler spent their active army careers as topographical engineers. During the 1820s, when West Point graduates were virtually America's only trained engineers, the government

Folk art lithograph of Russian railroad, Pesnia. Bliz Krasnykh Vorot. 1860s. (Print Collection, Miriam and Ira D. Wallach Division of Art, Prints and Photographs. The New York Public Library, Astor, Lenox, and Tilden Foundations)

began lending their services to private companies. McNeill and Whistler went to the Baltimore & Ohio Railroad. Other railroad and related engineering projects followed, both before and after they resigned their commissions, Whistler in 1833, McNeill in 1837. Whistler went to Russia in 1842 to supervise construction of the Moscow–St. Petersburg railroad, Russia's first, and he never returned.

Military-trained engineers came to comprise a significant share of midlevel managers. Furthermore, they instructed their civilian counterparts in the new methods. Like civil engineers, potential corporate managers benefited from the fruits of military education, whether in the form of on-the-job training in engineering projects or more formal schooling in colleges of engineering. No less important, the army offered models of large-scale organization and techniques for coordinating large workforces.

West Pointers adapted military methods to the new demands of railroad management. As the first large-scale business enterprises in America, the railroads in turn shaped the development of corporate organization in America throughout the nineteenth century. Management by staff and line, a key feature of the rising corporation, had self-evident military roots.

THE AMERICAN SYSTEM

Interchangeable parts manufacturing or the uniformity system, America's first major contribution to military-technological innovation, was a step removed from the field of battle. Its evolution spanned America's transition from a preindustrial to an industrial economy. The technique had French roots, like the military academy at West Point and many other aspects of military institutional development in the young American republic. The first firearms manufactured at the new Springfield Armory, for instance, simply copied a French model.

Uniformity cannot be so firmly pinned down. Beginning in 1763, Jean Baptiste Vacquette de Gribeauval tried to create standard guns and gear for the French artillery. His reforms provided the mobile field guns that Napoleon would so famously and decisively deploy in victory after victory. Gribeauval also promoted the work of Honoré Blanc, who from the mid-1780s sought to produce in French armories small arms with uniform parts. Developed further in U.S. Army arsenals during the first half of the nineteenth century, this approach decisively reshaped the organization and equipment of armies as the so-called American system of manufactures.

Though Eli Whitney was a key figure in the development of the American system, the reason was not, as most textbooks have it, because he invented a way to make muskets with interchangeable parts. Whitney was more promoter than inventor of the uniformity system, as it was then called. Imposing uniformity on musket parts would have allowed him to produce large numbers in relatively short order; on the strength of his promise to deliver 10,000 stands of arms in two years, he obtained in 1798 a War Department contract and a large cash advance, the first of its kind. Only in numbers did Whitney meet the terms of his contract. None of the muskets arrived on time, the last of them being nine years late, and their quality was poor.

Despite the shoddy performance of Whitney the arms maker, Whitney the manager introduced important innovations in the division of labor and cost accounting, and Whitney the publicist did much to promote the uniformity system. The concept was sound enough, as Gribeauval and Blanc had shown in France. Whitney learned of the French work from Thomas Jefferson, who witnessed some of their results when he was American minister to France in the late 1780s. Putting the system into practice was another matter. That would have taken more understanding, effort, and time than Whitney could muster. But he did become an ardent promoter of a still novel approach to arms making.

A more direct, if less well-known, source of French technological ideas was Louis de Tousard, a graduate of the French artillery school at Strasbourg who was seconded to the American army during the Revolutionary War. Returning to the United States in 1793, he became a key figure in creating the military academy at West Point. His three-volume *American Artillerist's Companion*, a standard text at West Point for years after its 1809 publication, strongly promoted the principles of uniformity and regularity that Gribeauval had pioneered.

Neither Whitney nor Tousard, however, played major roles in converting ideas to practice. Other private contractors did better. Simeon North and John H. Hall, in particular, possessed the insight and skill to create a true uniformity system. In this effort they were strongly encouraged by George Bomford during his two decades as army chief of ordnance. Created early in the century to replace a loosely supervised group of private suppliers as the army's primary source for artillery, firearms, and ammunition, the Army Ordnance Department also took charge of the nation's arsenals. Bomford became chief of ordnance in 1821. During the 1820s and 1830s, working closely with army arsenal managers, notably Roswell Lee at the Springfield Armory in Massachusetts, Hall and North laboriously assembled the pieces of a practical system of uniformity manufacturing: division of labor among the workers, specialized machinery, and precision methods of measurement. The U.S. Model 1842 musket, designed and manufactured at the Springfield Armory, was the first weapon produced in quantity with interchangeable parts.

Simeon North (1765–1852)

North's life changed when he opened a scythe-making business in an old mill next to his farm in Berlin, Connecticut. He may already have been making pistols for sale when, in 1799, he won his first government contract. It gave him a year to provide 500 pistols.

Simeon North. (Smithsonian Institution)

Additional contracts followed at regular intervals, and by 1813 he employed as many as fifty men.

In 1813 North took a contract for 20,000 pistols with interchangeable parts, a concept with which he had already been working for some years. He had also devised what may have been America's earliest milling machine, a key machine tool in interchangeable parts manufacturing. North added a new factory in Middletown, Connecticut, to his Berlin operation and continued to prosper. During the 1820s he switched from making pistols to manufacturing long arms, muskets, and carbines. Government contracts continued to flow to this factory until his death.

Methods and machines pioneered in the arsenals spread through the American arms industry, most famously to Samuel Colt, who based the manufacture of his patented revolving six-cylinder pistol on the uniformity system. Colt had struggled until the Mexican War brought army contracts. By 1851, when the Great Exhibition in London brought these novel techniques of mechanized production to the world's attention, they had already spread to other manufacturing enterprises. Colt's display at the 1851 exhibition proved a great success and he was flooded with orders. By 1855 he could open his own state-of-the-art armory using interchangeable parts manufacturing and assembly-line methods, a showplace for the American system of manufactures and a training ground for its transfer to other industries. During the Civil War the Colt Armory was second only to the federal Springfield Armory in the number of weapons produced.

EARLY NAVAL INNOVATION

Like the Revolutionary War, the War of 1812 that confirmed American independence saw the United States display technological novelties at sea. The most novel and least successful came from the indefatigable Robert Fulton. Long before he achieved fame by promoting commercial steamships, he designed and built a muscle-powered submarine for Emperor Napoleon of France. Launched in 1801, the *Nautilus* used ballast tanks to submerge—it had enough air to sustain a four-man crew for three hours—and a horizontal rudder to maintain depth. Despite successful trials, Fulton's boat proved ineffective in combat, and he had no better luck trying to sell his idea in England.

Robert Fulton (1765–1815)

Reared with four siblings by his widowed mother in Pennsylvania, Fulton learned the jeweler's trade in Philadelphia. After a four-year stint as a landscape painter, he left for England in 1786 and remained in Europe for the next two decades. Among his patented inventions were a double inclined plane for raising and lowering canal boats, a powered machine for dredging canal channels, and a prize-winning machine for sawing marble.

His attempt to interest first the French, then the British, government in his designs for self-propelled torpedoes and an underwater

Submarine above and below water level, 1806 pastel
drawing by Robert Fulton. (Library of Congress)

boat proved unsuccessful, largely because the devices themselves
failed to work as advertised. But his contact with Robert Livingston,
minister to France and later governor of New York, proved more
fruitful.

Not only did Fulton later (in 1808) marry Livingston's cousin,
Harriett Livingston, but also contracted with him for a commercial
steamboat service on the Hudson River, which opened in 1807. The
U.S. Congress finally provided the government support he had long
sought, first for further torpedo experiments, then for construction of
the steam-powered warship *Demologos*.

But in the United States, he persuaded Congress to fund an even more
ambitious project, a steam-powered vessel with a hundred-man crew. It
was launched in 1814 but remained untried when its inventor died. Fast,
well-armed, nearly impervious to existing ordnance, *Demologos* (also known
as *Fulton I*) completed her trial runs only after the War of 1812 had ended,
and so never entered service. Unseaworthy and intended strictly for harbor
defense, she inspired little enthusiasm among naval officers. Many years
passed before steam warships again appeared on the scene.

Another naval innovation, this one the result of improving tradition
rather than promoting radical change, enjoyed greater success. The Naval
Act of 1794 produced three famous frigates, the *Constitution* and *United*

States, each of 44 guns, and the 36-gun *Constellation*, all launched in 1797. They outgunned the British frigates they fought in the War of 1812, and outsailed the men-of-war that might have overwhelmed them. Although none of this much affected the outcome of the war, their exploits did help create an American naval tradition. Sail-powered warships, however, were nearing the end of the line. The future belonged to steam, even if few yet recognized the fact.

Naval innovation encompassed science as well as ships. Matthew Fontaine Maury and Alexander Dallas Bache emerged from naval backgrounds to become leading early-nineteenth-century scientists. Maury started as a midshipman in 1825, but when an 1839 accident confined him to shore duty, he had already made a name for himself. In 1842 he was appointed superintendent of the depot of charts and instruments in the Navy Department, a position he held until 1855, then again from 1858 to 1861. His work on wind and current charts founded modern oceanography. He also directed the Naval Observatory from 1844 to the outbreak of the Civil War. Maury served in the Confederate navy throughout the Civil War as director of coast, river, and harbor defenses, a position in which he encouraged such innovations as electric mines. He fled to Mexico after the war and spent some time in England, but returned to America in 1868 to become professor of meteorology at Virginia Military Institute.

Bache made his mark with the United States Coast Survey. After graduating from West Point and serving four years in the Corps of Engineers, Bache had resigned his commission to begin teaching natural philosophy and chemistry at the University of Pennsylvania. His academic career lasted into the 1840s, but was increasingly overshadowed after 1843 by his role as superintendent of the United States Coast Survey, a post he held until his death, and by his active promotion of science. He was a founder of the American Association for the Advancement of Science, an incorporator and regent of the Smithsonian Institution, a founder and first president of the National Academy of Sciences, and a president of the American Philosophical Society. During the Civil War he served as vice president of the Sanitary Commission.

Besides its titular function, the Coast Survey pioneered studies of geodesy and hydrography. Although a civilian agency, the survey made extensive use of naval officers. Benefits flowed both ways. The survey enjoyed a pool of able men, while officers gained valuable experience and knowledge that the small active navy of those years could not have provided them. A disproportionately large number of senior naval officers in the Civil War had honed their seafaring skills in the antebellum Coast Survey.

MILITARY AND NAVAL EXPLORATION

During the early nineteenth century, American science remained more strongly oriented toward observation than experimentation. The vast and little-known continent stretching westward reinforced that bent. Crossing the next hill seemed always to promise a strange plant or animal, some exotic tribe, an unexpected lay of the land. A large if not entirely constant stream of specimens and notes, drawings and photographs, maps and descriptions from western exploration promoted vigorous American contributions to such primarily observational fields as astronomy, natural history, the earth sciences, and anthropology.

Even when explicitly designated scientific, however, western expeditions served other purposes as well. Identifying potential resources for exploitation was an obvious end. So, too, was the quest for information of military value. Virtually every scientific expedition westward during the early and middle nineteenth century was also a military reconnaissance. The army played a key role in exploring the vast new lands acquired by the nation through purchase (such as Louisiana and the watersheds of the Mississippi and Missouri rivers, 1803) and conquest (such as California and the Southwest, 1848) during the first half of the nineteenth century. Military organization and discipline proved invaluable in exploring a little-known and sometimes hostile country, as did the skills of the engineer. Mapmaking and fort building along the way were normal activities.

The pattern was set from the beginning, when President Thomas Jefferson in 1803 selected army captain Meriwether Lewis to explore the vast new territory of Louisiana purchased from France. Lewis invited army veteran William Clark to share command of the forty-man expedition, which took just under two and a half years to cross the continent from St. Louis to the mouth of the Columbia River and return. Publishing the results consumed far more time than obtaining them. During the next three decades, other military-scientific expeditions penetrated western lands, surveying, mapping, and collecting data.

Prominent among the military explorers were members of the U.S. Army Corps of Engineers' Topographical Bureau, established during the War of 1812. Activity accelerated toward midcentury, especially after 1838 when the bureau became the Corps of Topographical Engineers. Unlike the regular army engineers who worked mainly on construction and fortification, the topogs, as they were often called, specialized in mapping and surveying. Their skills and hard work opened lands formerly known only to native inhabitants and a relatively small number of fur trappers and traders to economic exploitation and a growing influx of settlers from the eastern United States and from Europe.

Under the leadership of Col. John James Abert, the topog efforts were usually justified in such practical economic terms as surveying routes for a transcontinental railroad, but actual projects rarely had so narrow a focus, and scientists assumed larger roles in planning and manning western expeditions. During the three decades Abert headed the corps of thirty-six officers, most of them also West Pointers, they helped turn a relatively unknown wilderness west of the Mississippi into an extensively surveyed, well-mapped, built-up, interconnected, and increasingly settled landscape.

As a new lieutenant in the topogs during the 1850s, Gouvenour Kemble Warren participated in surveys of the Mississippi Delta and railroad routes to the Pacific. He later compiled the first accurate map of the trans-Mississippi West. A lifelong champion of Indian rights, Warren in the mid-1850s amassed an important collection of Sioux and other northern Plains Indian material culture. He sent back crates of Native American artifacts and natural history specimens to the fledgling Smithsonian Institution. Warren also made extensive geological and meteorological observations, duly reported to scientists back East. His contributions joined the stream of specimens and notes from western exploration upon which such sciences as natural history, geology, and ethnology flourished in America.

Midcentury also saw the U.S. Navy assume a major role in exploration alongside the army. Although motives again were mixed, science played a larger part in the naval enterprise and the geographical spread was far greater, ranging from the Arctic to the Antarctic, from the New World to the Old. Lt. Charles Wilkes commanded the most famous, the United States Exploring Expedition of 1838–1842. Publishing the expedition's results and observations entailed a struggle ten times longer than the expedition itself.

Reciprocity of benefit was not the normal outcome of the military-scientific partnership in early-nineteenth-century America. Association probably benefited American science far more than it did either army or navy. Scientists busy collecting, cataloging, and describing had as yet little to offer the armed forces. Much the same was true of engineering. Military engineers fostered engineering education, helped build the nation's infrastructure, introduced major features of large-scale management, and pioneered mass production—all of immense value to economic growth. Reciprocal benefits of any direct nature scarcely existed.

Indirect effects, however, proved enormously significant when the southern states seceded from the Union. Economically, politically, and socially, the nation that divided in 1861 was very different from the republic formed in 1789, or even the one that emerged from war in 1815. Such changes in part reflected the interactions of military institutions with American society during the preceding six decades, and they strongly influenced the nature and course of the war.

2

Transition to the Industrial Age: Mid-Nineteenth Century

◆

Before the nineteenth century, the complex interrelationships among war, science, technology, and industry so characteristic of the modern age scarcely existed, even in embryo. The French Revolutionary and Napoleonic wars had seen the triumph of mass armies, but technologically they belonged to the past, not the future. That changed in the middle third of the nineteenth century, in the United States no less than Europe, as industrialization and military-technological innovation began to exert far-reaching effects on the conduct of war. Practical rifled firearms and steam locomotives dramatically transformed land combat. At sea, steam power and armor worked a no less dramatic transformation, marking the first stages of the nineteenth-century naval revolution. Occurring in the midst of these changes, the American Civil War was notably transitional. Commanders looked to the past for tactics and organization, even as battle itself presaged the growing mechanization of war.

The United States entered the first stages of the nineteenth-century military-technological revolution by the 1830s. Change far more rapid than ever prevailed under preindustrial conditions became the rule. More telling, the rate of change increased. As time between idea and product grew ever shorter, the pace of meaningful military-technological change accelerated. Craft traditions gave way to factory production, random invention to planned innovation, hit-or-miss improvement to organized research and development. In time the progressive and rapid improvement of old, as well as the introduction of new, weapons would become core factors in military planning and operations, but that took decades. Engineering and industry had yet to attain the scope and mastery that allowed the speedy development of warlike inventions and improved weapons, or their prompt production in quantity. Although that ability had scarcely emerged before World War I, nor become fully realized until World War II, its prospect could be glimpsed even during the course of transition.

The American Civil War straddled two ages. It was both the last great preindustrial war and the first major war of the industrial age. On the eve of war the United States was not yet an industrial nation, but the process of industrialization was well under way. A key factor was steam power applied to produce and transport goods. Military engineers had once again assumed a prominent role. West Point provided many of the railroad builders, and Annapolis later began to bolster the ranks of steam engineers. Industrial capacity attained new levels of military significance as transportation improved, but in this, as in many other respects, the Civil War was distinctly transitional, and that strongly affected its conduct.

TOWARD MODERN SMALL ARMS

Springfield Armory's Model 1842 smoothbore musket, first product of the new uniformity system from a government arsenal, replaced the flintlock with the new percussion caplock. Caplocks eliminated the external ignition that made flintlocks prone to misfire. The cap, attached to the weapon through a sealed port, held an unstable chemical compound; a sharp blow from the hammer detonated the chemical, sending a flame through the port to ignite the gunpowder in the chamber. Percussion ignition greatly enhanced reliability. When allied with a practical rifling system, as it soon was, it helped radically transform small arms.

Rifles had existed at least since the fifteenth century. Unquestionably superior to smoothbores in range and accuracy, they suffered from one major drawback for military purposes. Bullets had to fit the barrel tightly for rifling to work. Ramming a tight-fitting bullet down a three- or four-foot barrel, as muzzleloading rifles required, was a hard, slow job, even compared to the none-too-easy task of loading a smoothbore musket. One answer was breechloading, which, like rifling, had a long history and a serious military drawback: gas escaping from the breech. Only in the nineteenth century did manufacturing techniques attain precision enough to meet military needs for cheap, plentiful, and reliable breechloaders.

One stopgap answer to the rifling problem came to fruition in the 1840s, the brainchild of French Army Capt. Claude-Étienne Minié building on earlier work by another French army inventor, Gustav Delvigne. The key idea was a bullet small enough to slip easily down the barrel, which could then be deformed to grip the rifling. Delvigne proposed tapping ball with ramrod to deform the bullet, but Minié suggested something more elegant, the self-expanding bullet. Exploding powder would expand the bullet's hollow base, spreading the lead to fill the barrel and grip the rifling. Although Minié's name became firmly attached to the new bullet, the actual conversion of Minié's largely theoretical solution into a practical device was the handiwork of an American, the assistant master armorer at the army's Harpers Ferry Armory, James Henry Burton.

The U.S. Model 1855 rifle-musket incorporated this system, plus an improved percussion system. Earlier model smoothbores were also converted to rifles, and percussion rifled muzzleloaders became the mainstay of Civil War infantry. Far more accurate and dependable than their predecessors, they also outranged the smoothbore gun that still formed the backbone of army artillery. When the technical framework for the tactics that defined Napoleonic warfare ceased to exist, consequences were written in blood on every Civil War field. By extending severalfold the zone of fire through which attackers must pass, rifled muskets swung the tactical balance back toward defense. Smoothbore artillery with an effective range of 400 yards far outranged smoothbore muskets. On Napoleonic battlefields massed artillery allowed attackers to decimate a defending force, clearing the way for a decisive bayonet assault. Rifled small arms reversed the advantage in the 1860s. Defenders entrenched or sheltered behind breastworks beyond the reach of smoothbore cannons could wreak havoc on attackers crossing open ground.

Under the new circumstances, frontal assault could succeed only at terrible cost, and bayonets hardly mattered. Yet Civil War commanders persisted in ordering mass assaults with usually disastrous results. Given the chance, soldiers quickly learned to dig, the spade becoming little less important than the rifle. Before it ended, the Civil War had become an engineer's war. Field fortifications, often neglected early in the war, soon became normal. Troops began disappearing into the ground or behind breastworks at even brief pauses. By the final year, elaborate trench systems converted field operations into siege warfare. In this as in so much else, the Civil War in America foreshadowed war to come.

Breechloading, the next crucial step in small-arms development, did not much affect Civil War combat. Breechloading weapons entered service well before metal cartridges finally solved the problem for good. The same

John Hall who helped perfect the American system also invented the best flintlock breechloader, adopted by the U.S. Army in 1819. The link was not accidental. Hall received not only a contract specifying royalties on his musket, but also a job at the Harpers Ferry Armory where he helped advance the uniformity system while working to improve his design. A clever solution to the problem of leaking gas allowed Hall's protégé, Christian Sharps, to devise the Sharps Model 1859, the last word in percussion breechloaders. Though Sharps' rifles saw wide service, rifled muskets dominated the battlefields of the Civil War. Only after the war and the development of practical metal cartridges did breechloading fully replace muzzleloading.

John Hancock Hall (1781–1841)

Stephen Hall and Mary Cotton Hall of Portland, Maine, both from prominent New England families, had six children. The oldest, John, showed little interest in following his father and grandfather to Harvard Divinity School. He preferred invention to scholarly theology. As a young man serving in a Portland militia company, he saw ways that the firearms carried by citizen soldiers might be

Open breech of Hall rifle. (Smithsonian Institution)

improved. In 1811 Hall received his first patent for a breechloading system. In 1819 a government contract brought him to Harpers Ferry, Virginia, where he produced a test lot of breechloaders tooled under a new system with parts that could be interchanged. From 1820 to 1840, Hall worked assiduously on breechloading firearms and interchangeable parts manufacturing—the uniformity system— helping to perfect America's first major contribution to industrial development.

ORDNANCE IN TRANSITION

Attempts to apply rifling to artillery through the mid-nineteenth century met less success. Metallurgical practice could not yet produce ordnance strong enough to withstand the pressure of a fully contained explosion, except in relatively small caliber guns. Metal barrels too often burst under the strain of explosions confined by tight-fitting projectiles, unless firing reduced powder charges. Unfortunately, smaller charges meant lesser impact and limited effectiveness, even for the best rifled artillery of the mid-nineteenth century. Such guns lacked killing power, leaving rifled small arms master of the field. West Point graduate Thomas J. Rodman developed a new founding process and introduced a reasonably satisfactory 3-inch rifled gun, but the most widely used rifled field gun in the Civil War was the work of another West Point graduate, Robert P. Parrott.

Although rifled ordnance had begun to assume growing importance on midcentury battlefields, improved smoothbore guns firing shells and shrapnel as well as solid shot played the larger role into the 1860s. Outstanding among these weapons was the 12-pounder field gun of 1857, called the "Napoleon." Named after the chief author of the improved design, Louis Napoleon (the sometime Emperor Napoleon III of France who had devoted extensive study to artillery of all kinds before he ascended the throne), it may have been the best muzzleloading smoothbore cast-metal field gun ever made. It was also the last. In a pattern familiar to historians of technology, an obsolescent technology achieved its finest expression just before its demise.

Considerable progress marked midcentury efforts to regularize and systematize the army's artillery along the lines that Tousard had championed earlier in the century. West Point graduate Alfred Mordecai advanced the cause after he transferred to the Ordnance Corps in 1832, and played a major role in reorganizing army artillery along more rational lines.

For three decades he promoted the application of scientific methods to developing and testing weapons and ammunition.

Alfred Mordecai (1804–1887)

Mordecai led his West Point class (1823) academically and taught engineering at West Point for two years. In 1832 he shifted to the Ordnance Corps in which he pioneered the application of scientific methods to developing and testing weapons and ammunition. He also played major roles in compiling the army's first ordnance manual (1841) and reorganizing army artillery along more rational lines (1849).

In 1861 Mordecai (a North Carolinian) resigned his commission, refusing to break his oath but unwilling to fight against Southern family and friends. He spent the war years teaching math in Philadelphia, near the family of his wife, Sara Ann Hays Mordecai. Their son Alfred graduated from West Point in 1861 and fought for the Union. After the war, Mordecai turned to railroading, briefly as

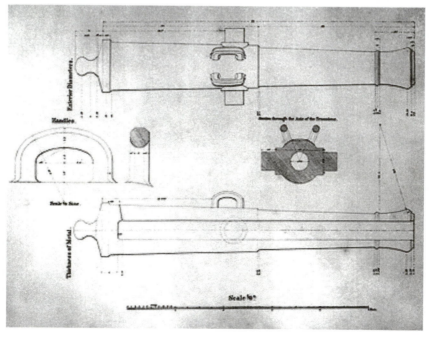

Plate drawing of a field 12-pounder bronze gun. (Smithsonian Institution)

an engineer in Mexico, then as an official with the Pennsylvania Railroad.

Naval ordnance likewise remained predominantly smoothbore, but enjoyed the first notable advances in two centuries. Stimulated by Henri-Joseph Paixhans, a French artillery officer, especially his 1822 *Nouvelle force maritime*, European navies began to adopt explosive shells of improved design in the 1830s. Eventually, the United States followed suit. Because shells were hollow, they could be larger than solid shot of comparable weight. Paixhans also advocated standardization for naval guns, another innovation soon under way. A Swedish-American naval officer and inventor, John A. B. Dahlgren, learned how to raise the caliber of shell guns from 8 to 11 inches, and Rodman's new founding process made guns with calibers up to 20 inches feasible. Explosive shells ended the day of steam warships propelled by paddle wheel. Improved naval guns also promoted the use of armor on ships.

THE NAVAL REVOLUTION EXPANDS

After Fulton's turn-of-the-century experimental steam warship failed to win a place, the U.S. Navy abandoned steam for decades. Abroad, steam warships remained minor, specialized fleet adjuncts. Steam power afloat posed several intractable problems in the early nineteenth century. Machinery remained bulky and inefficient. Installed above the water line to push paddle wheels, it also remained far too exposed for a warship, especially when shell guns appeared. Paddle boxes meant fewer guns, thus lesser weight of broadside. Fuel demands that sharply constrained cruising radius only made matters worse.

At a single stroke, screw propulsion seemed to resolve most of the problems. One of the first three U.S. Navy vessels mounting shell guns, the 10-gun sloop *Princeton* launched in 1843, was also the world's first warship driven by a stern-mounted screw propeller. Designed by transplanted Swedish engineer John Ericsson, who also oversaw its building, the 400-horsepower engine turned a six-bladed screw that gave *Princeton* a top speed of 13 knots.

The extreme vulnerability of wooden ships to shell guns also promoted the next major changes in naval architecture, iron hulls and armor plate. Here again the United States was a pioneer, when Congress in 1842 authorized construction of an all-iron armorclad war steamer. Like *Demologos* an essentially private venture the navy endorsed only reluctantly, the so-called Stevens Battery (after its builders, the brothers Robert L. and Edwin A. Stevens) suffered from bureaucratic delays and rapidly changing technology. Ordnance growing in power outstripped efforts to armor the ship against penetration by shot or shell. Despite three decades of work, she was never completed, and eventually sold for scrap.

When the Civil War began, the Federal navy had eighteen relatively modern steam-powered, screw-propelled, wooden warships. Iron hulls and armor had yet to enter regular service. Although steam proved indispensable to Federal blockade of the Confederacy, iron and armor were not. The burden of blockade fell on wooden ships, which were built at a furious pace. Armorclads played only an auxiliary role, though an important one, dictated in part by the South's reliance on makeshift armorclads to defend its ports.

The Confederacy enjoyed a cadre of Annapolis-trained officers, but began the war without ships. It also lacked factories for armor plate and marine engines, hampering naval defense throughout the war. Yet Confederate efforts to rectify the imbalance produced the war's most famous naval action and its best-known military innovation. Despite (or perhaps because of) its indecisive outcome, the 1862 clash off the southern Virginia coast in Hampton Roads between the Confederate *Virginia* (better known as *Merrimac*, the name of the ship upon whose hull she had been constructed) and the Federal *Monitor* marked a historic moment. As the first meeting between steam-driven, ironclad warships, it dramatized as nothing else could the revolutionary developments in naval technology then under way.

Of the two, *Monitor* was by far the most novel, another product of the inventive John Ericsson. Nearly everything about her was new and untried, but she became the namesake for an entire class of warships that persisted for decades. Ericsson himself made the detailed drawings that guided the workmen as he went along. Upon *Monitor*'s 124-foot hull was riveted a raft-like deck 172 feet long and 41.5 feet at its widest, with 4.5-inch iron armor on her sides, 1 inch on her deck. Most distinctive was the 140-ton turret amidships, a cylinder 9 feet high and 20 feet across encased in 8-inch armor. Revolving on a spindle cogged to an auxiliary steam engine, the turret could swing two 11-inch Dahlgren guns through a full 360 degrees, a field of fire unblocked by masts, rigging, or other sailing-vessel paraphernalia. *Monitor* relied entirely on her engines, also of Ericsson's design, two decades before most warships gave up their auxiliary sails.

John Ericsson (1803–1889)

Born in Sweden and home-schooled, Ericsson later trained in the Swedish Corps of Mechanical Engineers and served in the Swedish army for six years as a topographical surveyor. An invitation from Robert F. Stockton, American naval officer and influential politician, brought him to the United States in 1839. His wife of three years, Amelia Byram, never moved permanently to the United States, where Ericsson would spend the rest of his life.

Ericsson's numerous innovations improved power transmission for marine engines, screw propulsion, ship design, and ordnance. His screw propeller for boats operating on American inland waters proved an immediate success. Less happy was his collaboration with Stockton on the first screw-propelled warship, the USS *Princeton*. The problem was not the ship, but a gun of novel design it carried that exploded during its 1843 trials. Many died and Ericsson's reputation suffered, though Stockton had taken most of the credit for the innovative design. Ericsson's greatest triumph came in the American Civil War, when he brought his revolutionary design for an armored, screw-propelled warship with a revolving turret from plans to combat in only a hundred days.

USS *Monitor* crewmen cooking. (U.S. Navy)

Virginia, in contrast, was built on the remains of a standard wooden hull, armor merely draped over a wooden casemate housing two pivoted 7-inch rifles, plus three 9-inch Dahlgren smoothbores and two 6-inch rifles on each broadside. She also carried a heavy iron wedge fixed to her bow as a ram, largely useless because her rebuilt engines could manage only 4 knots. Heavily damaged in the battle, she required extensive repairs, then had to be destroyed when Federal forces seized her base at Norfolk, Virginia.

Virginia's guns and armor came from Tredegar Iron Works in Richmond, Virginia, a company owned by West Point graduate Joseph Reid Anderson. By 1860 the foundry was one of the nation's largest, producing locomotives, boilers, cables, naval hardware, and cannon. When war came, it emerged as the industrial heart of the Confederacy, using slave and free labor. Under Anderson's supervision, Tredegar remained the South's premier source of armor and heavy ordnance throughout the Civil War. Its vital importance to the Confederate war effort made Richmond's defense as much a military and economic as a political necessity.

Unsolved design problems largely precluded the use of oceangoing ironclads by either side. Notoriously unseaworthy, they remained restricted to protected coastal waters. Such shortcomings mattered little on Western rivers, where armored gunboats played a key role in the federal conquest of the Mississippi. Steamboats had multiplied on American inland waters before the war, having solved some of the worst problems of river navigation, especially in the West. Fitted with iron plate and guns, they became valuable adjuncts to army operations.

STEAM POWER AND LOGISTICS

Transatlantic and coastal steam-powered vessels had become almost commonplace by the 1860s. Potentially, steam made the resources of Europe more readily available to both sides. Practically, it made the federal blockade of the 3,000-mile confederate coastline feasible. Steamers could patrol waters off southern ports more closely and more regularly than sailing ships ever could have. The price was frequent recoaling. Federal forces seized bases on the Confederacy's coast as much to maintain the blockade as to deny their use to the enemy. Armorclad warships proved invaluable in these operations.

On land, steam-powered transport had far-reaching consequences for American development during the first half of the nineteenth century. Railroads offered a degree of flexibility that waterways could never match.

Here the prewar changes were revolutionary. In 1830 the Baltimore & Ohio Railroad put 13 miles of track into service. Competing and expanding lines laid over 9,000 miles of track in the next two decades, mostly in the Northeast. Between 1850 and 1860, another 22,000 miles of track extended the rail net to the South and West, though the Northeast retained the densest network. When the Civil War began, over 31,000 miles of track crisscrossed the country east of the Mississippi.

Overenthusiastic from an economic viewpoint, prewar railroad-building acquired strategic significance with the outbreak of war. Understanding fully what that meant, however, took time. Unlike at least some Europeans, Americans had given little if any thought to rail's military implications. In brief, railroads allowed contending states to marshal their resources swiftly. Large bodies of troops might quickly be shifted to where they were wanted. Perhaps more important, steam-powered logistics promised to maintain much larger armies than would otherwise have been feasible.

Speed, both of troop movement and of supply, emerged as the essence of rail's military utility. A defeated army falling back on its railheads might soon be repaired, its casualties replaced, its losses made good. Reinforced and reequipped armies could return to fight again, so long as men and materiel were available. Rapidly restored armies robbed battle of its potential for prompt decision. The two Civil War battles most often called decisive, Gettysburg and Vicksburg, were both fought almost as near the war's beginning as its end. They were also greatly prolonged. Vicksburg, of course, was a siege, so its length would not be surprising. But Gettysburg was a single battle lasting three days, a little-noted portent of things to come. Ultimately, one side's exhaustion decided the issue, and that would become true not only of battle, but also of war.

Dramatically increased speed of movement came at the cost of freedom of movement. Civil War armies became more and more closely tied to their railroad supply lines, the attack and defense of which came increasingly to dominate strategy. Building and wrecking rail lines became a major military function, but most railroad operations remained almost exclusively profit-making ventures in private hands. It was not an easy relationship. Railroad managers professing concern for their shareholders and army officers demanding logistic support were often at loggerheads. Although the North succeeded better than the South in imposing a degree of order, the South's real problems lay elsewhere. In this as in so much else, materiel shortages hampered the Confederate war effort: too little track, not enough rolling stock, and insufficient means to produce more.

Militarily, railroads mattered chiefly to the final element in logistics, distribution. Equally important, though hardly clear at first to planners on either side, was production. Here, too, steam power had already begun to work massive changes in manufacturing organization before the war. In due course the patent importance of rails led eyes back to the supplies, materiel, and equipment they carried forward to the fighting troops. Eventually, this meant that war enlarged its scope to include not only the armies, but also the resources and people that supported and sustained them. Sherman's trail of destruction through Georgia and the Carolinas or Sheridan's devastation of the Shenandoah Valley prefigured a new kind of war. Yet in this, as in many other ways, the Civil War was distinctly transitional—a surprisingly active intersectional trade, for instance, flourished throughout the war.

TECHNOLOGY, SCIENCE, AND MEDICINE

The Civil War in America was the first full-scale war shaped in major ways by the tools and weapons of the Industrial Revolution. Telegraph and railroad greatly increased the pace at which events moved, and combined with other technical changes to vastly extend the scope and deadliness of battle, while at the same time reducing its decisiveness. New weapons multiplied the ranges at which death could be dealt, but other factors multiplied the numbers who might be killed: Agricultural mechanization permitted larger armies to be fed; industrial growth, to be armed and supplied; steam-powered transport, to be deployed and sustained.

Science played little part in these changes. Although created by Congress during the war, at least partly as a source of advice on war-related matters, the National Academy of Sciences achieved little of import. Explored as a source of new weapons, applied research met little success. Unimpressive as the results were in the actual course of the Civil War, the idea triumphed. Memory tended to invest the efforts of the National Academy of Sciences with more success than they had actually earned. Many years elapsed, in fact, before it became a major actor in American science, but when the United States once again entered a major war such efforts were promptly resumed.

In contrast to more recent experience, the Civil War did more to disrupt than to advance science. Although not fully evident at the time, the long-standing military support of science through exploration was one of the casualties of the Civil War. West Point ceased being exclusively an engineering school, the Corps of Topographical Engineers expired in 1863, and the postwar Corps of Engineers itself became increasingly an executor of civil engineering projects conceived elsewhere.

Ingenuity and imagination, rather than science, marked the efforts of both sides to devise and apply new weapons and techniques. Civil War firsts—first used, or first used extensively—covered a broad range: mass production in some industries, notably clothing, and new techniques of food preservation; supplying and moving mass armies by railroad and steamboat; photography, telegraphy and various signal devices using flags and lamps, and aerial observation from fixed balloons; the general use of rifled smallarms and the appearance of breechloading and magazine arms, as well as early machine-gun systems; the normal disappearance of troops behind breastworks and into trenches, along with the use of wire entanglements and trench mortars; land and marine mines, torpedo boats, and submarines; and steam-powered armored warships. That arms production also included swords, lances, and pikes again attests the transitional nature of the American Civil War.

Where the military-technological innovations that began to transform warfare by the mid-nineteenth century actually came from still followed the age-old path from lone inventor through cut-and-try development. Team research was an invention of the later nineteenth century, and then only in industry. Organized military research took several decades longer. Nothing new significantly affected the course of the Civil War. The likeliest explanation is that industrial capacity had not yet reached the point where an innovation could be quickly and efficiently placed in large-scale production so as to equip very large armies. That would change decisively in the next half-century.

Medical care reflected another kind of transition. Again, a number of firsts may be cited, notably anesthesia in front-line surgery, establishment of an ambulance service, and use of hospital trains. The importance of sanitation and camp hygiene was widely respected, if not always fully implemented. One of the key figures in stressing the importance of hygiene in controlling the diseases that normally afflicted armies was America's first female doctor, Elizabeth Blackwell. Hygiene became a special focus of the United States Sanitary Commission, a civilian forerunner of the Red Cross that advised and assisted the Union armies. For all that, the Civil War ended just before medicine decisively crossed the line to science: General acceptance of microorganisms as the cause of disease and thus recognition of the value of antisepsis occurred in the 1870s. Improved sanitation and medical care nonetheless sharply altered the ratio of battle-caused to disease-caused deaths. As recently as the Mexican War, that ratio was still 1 to 10, the same that had prevailed in the Revolutionary War. In the Civil War, Union armies achieved a ratio no worse than 1 to 2, and even Confederate armies under much less favorable conditions managed 1 to 3.

Elizabeth Blackwell (1821–1910)

Blackwell grew up in England and America amid a reform-minded family's liberal and progressive ideas. Intent on becoming a physician, she faced a self-described great moral struggle upon finding medical schools closed to women. But she persisted. Only in 1847, when a medical school in rural New York mistook her application for a joke, did she win a place. Taking full advantage, she secured her medical degree, then studied in England and France. During the next decade, Blackwell promoted higher health care standards for women and children, opening female training clinics and hospitals in the United States. In England she worked with Florence Nightingale to advance medical training for women.

Blackwell family. Elizabeth is standing on left. (Schlesinger Library, Radcliffe Institute, Harvard University)

The outbreak of the Civil War brought new challenges. Long convinced of the importance of sanitation and personal hygiene in fighting disease, she organized the Women's Central Association of Relief, forerunner of the United States Sanitary Commission. Throughout the war, Blackwell and her younger sister, Emily, also a physician, selected and trained highly qualified nurses for war work. After the war, they established medical schools to provide women with both formal training and clinical experience.

INNOVATION VERSUS MANUFACTURING

Proliferating inventions were not chiefly what mattered in the Civil War. Many were impractical, others not widely used. Technological innovation counts only to the extent that it places weapons in the hands of troops. Not until a new device or technique comes into widespread use will its impact be felt. Steam-powered transport was one such innovation that strongly affected the course of the Civil War. Only one other among the host of Civil War firsts exerted an equally profound effect, though on tactics more than strategy: The troops on both sides carried rifled firearms. That such a situation was even possible owes much to the uniformity system developed in army arsenals earlier in the century.

When the Civil War began, the U.S. Army's standard arm was the Model 1855 rifle musket. Its adoption came three decades after Springfield Armory had begun its pioneering development of the uniformity system. Before then, Springfield never managed to produce more than 5,000 muskets a year. At such production levels, even with several government and contract armories involved, arming the hundreds of thousands who fought in the American Civil War with new weapons would have been well nigh impossible.

The Confederate seizure of Harpers Ferry Armory left Springfield as the only government arsenal as war began. Its annual capacity by then stood at roughly 12,000 arms. Substantial purchases abroad marked the opening of the war, and private contracts later augmented domestic arms manufacture, but Springfield remained the Union's major source of rifled muskets under the leadership of Alexander Brydie Dyer. By 1865, in fact, it had become the world's largest arms factory with an annual capacity of better than 300,000. The Confederacy of necessity relied more on foreign supply, but nonetheless managed to build a creditable arms industry almost from nothing, aided by the machinery seized at Harpers Ferry. Most of the credit belonged to Josiah Gorgas, a Pennsylvanian who followed his Alabama-born wife into secession and worked miracles in organizing arms production. As chief of ordnance for the Confederacy, he built an extraordinary system of acquisition, manufacture, and distribution of arms and ammunition. Largely thanks to him, Confederate troops never lacked weapons, though they often ran short of everything else. At their peak in mid-1863, southern armories were producing 28,000 small arms a year, augmented by 7,000 from private sources.

Like the uniformity system, the rifled musket that dominated Civil War battlefields was in some respects a typical product of premodern patterns of technological innovation. Technology, whether civil or military, was chiefly empirical. Vast as the accumulation of technical knowledge had become by the mid-nineteenth century, it was still normally the product of hit-or-miss accident by craftsmen or tinkerers, laboriously augmented over many years, unevenly developed, and slow to spread. But again like the uniformity system, rifle development incorporated something more novel, a kind of systematic empiricism that began from the eighteenth century onward to accelerate the pace of change. This trend grew even stronger as the century advanced.

3

The Military-Technological Revolution: Late Nineteenth Century

◆

Both army and navy reverted to prewar levels and tasks within a few years after the Civil War. Policing the frontier and patrolling distant stations again became their main functions. A kind of enforced isolation from the rest of American society promoted military professionalization. The armed forces, like other professionalizing groups during the late nineteenth century, sought to make special schooling a prerequisite for entry and advancement. Military education dramatically expanded and improved, both for officers and men. New schools for officers offered postgraduate courses in military theory, policy, and practice. Technical courses for lower ranks gave basic training in operating and maintaining ever more complex weapons and equipment.

During the late nineteenth century, United States armed forces became chiefly observers and sometimes adopters of major changes in military technology and organization taking place in Europe. The innovative role played by U.S. armed forces in military-technological development through much of the nineteenth century declined because they became less receptive to innovation, or because they simply no longer enjoyed levels of funding that allowed them to sponsor research and development. Individual American inventors and entrepreneurs continued to spin off ideas and proposals, but found their warmest welcome in Europe. From the end of the Civil War until well into the twentieth century, the United States remained largely a borrower of weaponry developed abroad.

THE MAKING OF PROFESSIONAL
OFFICERS AND SCIENTISTS

By the late nineteenth century, advanced schooling had become a normal part of an officer's career, and the concept of military science had become commonplace. The link was no accident. Military science codified the underlying principles of war, which could thus more easily be taught in the classroom. It abstracted and systematized a body of esoteric knowledge suited to indoctrinating the nineteenth century's growing numbers of nontraditional candidates for officer status. Presumably lacking the genetic predisposition of their elite comrades, they needed concrete and readily reproducible examples: schematic maps all could see, war games all could play, rules all could memorize. The so-called principles of war, for instance, were a nineteenth-century innovation.

Few of these teaching aids were entirely new, but their use burgeoned during the late nineteenth century and became a staple of the twentieth. Paradoxically, military education also grew more complex and sophisticated as some of its subjects became oversimplified and standardized. The reoriented curricula of older schools in Europe and the United States added courses in strategy and policy to the familiar tactics and engineering. At new military schools founded for that very purpose, postgraduate training became first available, then a required prelude to higher command.

The School of Application for Cavalry and Infantry, founded at Fort Leavenworth, Kansas, in 1881, marked the U.S. Army's first venture into this area; the Naval War College was established three years later in Newport, Rhode Island. Strategic and other higher military studies in the United States proliferated in the context of turn-of-the-century reform movements that affected military as well as civil society. In the twentieth century such studies have become a central feature of advanced military education. All such courses and programs pointed toward professionalization, officers educated to wield sanctioned violence responsibly.

Eben Swift (1854–1938) and Arthur Lockwood Wagner (1853–1905)

Swift and Wagner were career army men, Swift having grown up in an army family with a strong attachment to military life and organization, while Wagner, of German descent, was thirteen when his father died, leaving his family destitute in a small Illinois town.

Both made it to West Point, graduating in the mid-1870s into an emergent reform-minded society where concepts of professionalism and specialized graduate education were beginning to create new disciplines. As freshly commissioned officers assigned to separate posts in Indian Territory, Swift and Wagner were strongly influenced by these modernizing trends.

Posted to the army's recently established School of Application for Cavalry and Infantry, at Fort Leavenworth, Kansas, in 1887, they together pioneered the army's postgraduate education system, introducing transformative ideas for training soldiers and educating officers. Swift's contributions included adapting training methods from much admired Prussian military manuals and establishing a standardized system for writing field orders; Wagner developed courses in military science and tactics for officers. Both later became deputy commandants at the school, and Swift served briefly as commandant.

Profound and rapid social change strongly colored, if it did not cause, professionalization, whether of armed forces or other corporate groups during the nineteenth century. Engineering shifted from practical training on building site or shop floor toward increased emphasis on formal college education as prerequisite to professional certification. Science likewise expanded its educational demands, reflected especially in the late nineteenth century innovation of graduate programs and doctoral degrees, and the burgeoning number of professional societies. Medicine, bolstered by the success of the bacteriological revolution, emerged from a host of competing creeds to become the preeminent health profession, a claim reinforced by turn-of-the-century reforms in medical education.

MILITARY-SCIENTIFIC INTERACTIONS IN THE LATE NINETEENTH CENTURY

The close prewar links between military and scientific enterprise tended to fade with the increasing professionalization of both science and the military. Military-scientific cooperation in exploration revived during the 1870s and 1880s, though of much lesser extent than before the war. Army officers and troops systematically surveyed large areas of the trans-Mississippi West, completing the work undertaken before the Civil War by the topographical engineers. Naval exploration also continued its focus in the post–Civil War period on the Arctic. As before the war, scientists regularly accompanied the expeditions, some of which ended disastrously. From 1867, when the United States purchased Alaska from Russia, the army administered and explored the new territory. The army also played a crucial role in managing the newly developing national park system from 1886 to 1918, and may have saved the national park idea in the process.

Established by Congress in 1860, the U.S. Army Signal Corps was the world's first specialized tactical military communications organization. Like so many other aspects of Civil War science and technology, signals proved decidedly transitional. Signal corpsmen used a new system of wigwagging flags (or torches at night) to pass orders on the battlefield. They did not restrict themselves to line-of-sight signaling happily. Throughout the war, the U.S. Military Telegraph, technically under the Quartermaster Department but actually a separate and largely independent civilian organization, thwarted every Signal Corps effort to employ electrical telegraphy. But the Military Telegraph ended with the war, leaving the postwar Signal Corps in charge of military electrical communication.

When Congress assigned the newly created Weather Service to the Signal Corps in 1870, telegraphy became central to the corps mission. Over the next two decades, the Signal Corps grew from a weather reporting service to a weather forecaster and then to a center of meteorological research. In the late nineteenth century, none of this seemed particularly military, either to the army itself or to Congress, which in 1890 transferred the Weather Service to the Department of Agriculture. Much the same fate befell the Geological Survey, the Coast Survey, and even the Naval Observatory, all of which had, for varying lengths of time, worked under military auspices, and all of which passed to purely civilian control. Although the Corps of Engineers retained its military structure and affiliation, its orientation and personnel became ever more civilian and its work increasingly centered on such civil engineering projects as dams and waterways.

With the loss of its weather mission, the Signal Corps returned to its roots. Electrical communication had burgeoned in the preceding two decades and the army, in this area as in weapons technology, trailed European developments. That began to change in the 1890s as the Signal Corps shared in the growing professionalization of the armed forces. Telephony and radio joined telegraphy as civilian technologies that could fruitfully serve military purposes. The highly successful military career of George Owen Squier, who added a Ph.D. in physics to his West Point credentials, demonstrated how advanced scientific training might now serve military purposes. In addition to his own highly creditable radio research, Squier was instrumental in persuading the army of the benefits of organized re-search. The civilian electrical and chemical industries, leading sectors of what has been termed the second industrial revolution of the late nine-teenth century, had learned that diverting some of their profits to support fundamental research paid long-term dividends. The radio research labo-ratory at Fort Monmouth, New Jersey, and the aviation research laboratory at Langley Field, Virginia, that Squier helped establish would in time be-come the military equivalents of such industrial research laboratories as those of General Electric and Du Pont.

George Owen Squier (1865–1934)

Squier followed graduation from West Point with advanced study at Baltimore's Johns Hopkins University, becoming in 1893 the army's first Ph.D. Forty years of electrical research, including several basic radio patents, won him membership in the National Academy of Sciences. In 1905 he founded the army's first signals school at Fort

Colonel Squier inspecting the laboratory. (NARA)

Leavenworth, Kansas, and later the army's radio research laboratory at
Fort Monmouth, New Jersey. He also pioneered army aviation,
writing the first specifications for a military airplane in 1907 and later
founding the army's aviation research laboratory at Langley Field,
Virginia. Appointed Chief Signal Officer in 1917, Squier oversaw
all army communications and aviation in World War I. Although his
reputation took a bruising from wartime problems with aircraft
production over which he had little control, he deserves great credit
for institutionalizing research in the army.

PERFECTING SMALL ARMS

Military small arms continued their revolutionary changes in the late
nineteenth century. Breechloading, fixed ammunition, magazines, and
smokeless powder, all achieved practical form in rapid succession and
transformed the rifled musket of the 1860s into the modern rifle by the
1890s. Several forms of fixed ammunition incorporating primer, charge,
and bullet in a metal cartridge were devised in the middle years of the
nineteenth century. Such ammunition solved the chief problem of breechloaders, leaking gas, since the metal casing served as an effective obturating
agent. With improved manufacturing techniques and metal cartridges, arms
makers could meet military needs for cheap, plentiful, and reliable breechloaders.

Once metal cartridges and breechloading became standard, magazine rifles quickly replaced single-shot arms. The centuries-old concept
of multiple-shot weapons sprang from obvious advantages to soldiers who
could fire several times without reloading. Efforts to turn concept into
practical weapon, however, demanded breechloading. Metal cartridges not
only solved that problem, but their easy handling quickly led to magazines
and other repeating mechanisms.

Repeating rifles saw Civil War service, Christopher M. Spencer's by far
the most important. Connecticut born and bred, he became a machinist
and began working on a repeating rifle in the mid-1850s. The rifle he
patented in 1860 reached full production by 1863. Lever action rotated the
breechblock, inserting into the chamber a copper rimfire cartridge from
the seven-cartridge, spring-fed tubular magazine in the rifle butt. It proved
both effective and reliable. Technically the most advanced repeating rifle of
its time, it became the standard arm of Union cavalry by war's end. Spencer
received a one-dollar royalty for each rifle and carbine sold, 107,000 in all.

No repeating rifle, however, gained wide use; they were more portent of things to come than decisive influence on the war itself.

Two innovations completed the transformation of infantry small arms, metal-jacketed bullets and smokeless powder. The work of Swiss Majors Bode and Rubin resulted in the first successful metal-jacketed bullet in the early 1880s. Working in an industrial research laboratory, French chemist Paul M. E. Vielle invented smokeless powder in 1886 when he plasticized nitrocellulose with ether and alcohol. Smokeless powder suitable for firearms had two major consequences. First, it reduced to mere traces the formerly dense cloud of smoke that betrayed the soldier's position and obscured his vision. Second, and more important, its relatively slow-burning qualities yielded much higher muzzle velocities without matching increases in internal pressure. Because metal jackets prevented rifling from stripping or heat from melting the soft lead bullet, this innovation likewise contributed to sharp increases in internal velocity.

Beginning with the French Lebel in 1886, new small-bore rifles combined these changes to produce radically higher velocities, flatter trajectories, and increased range. When the U.S. Army in 1892 decided to replace the twenty-year-old single-shot Springfield with a new .30-caliber magazine rifle firing smokeless cartridges, it turned to the Danish Krag-Jorgensen, a knockoff of Paul Mauser's German magazine rifle. Not until 1903 did the United States design and produce its own magazine rifle, the bolt-action Springfield that remained in service until the eve of World War II.

MULTIPLYING FIRE

Even as soldiers' individual weapons achieved levels of efficiency that left them largely unchanged for the next half-century, new weapons tended to reduce their significance. Machine guns and quick-firing artillery dominated the second stage of industrial war. The machine gun had numerous forerunners, none of them truly ancestral. Early attempts to achieve rapid fire relied on multiple barrels. In this form the idea can be traced to the earliest years of gunpowder weapons. Workable systems, however, waited until the last half of the nineteenth century, when reliable fixed ammunition became available.

Richard J. Gatling patented the most successful of all multiple-barrel guns in 1862. Civil War models used copper rim-fire cartridges, gravity feed, and six rotating barrels. In the course of its long career the Gatling gun underwent numerous modifications, but its essential feature always remained the crank-controlled revolution of several rifled barrels about a central axis, each barrel firing as it came into contact with the breech mechanism. It remained the only authorized machine gun in U.S. service until 1904.

Richard Jordan Gatling (1818–1903)

Gatling was one of seven children born to a prosperous North Carolina farmer and inventor, Jordan Gatling, and his wife, Mary Barnes Gatling. Sharing his father's mechanical bent and inventiveness, the young man followed high school graduation with a variety of jobs while he worked on his own inventions. At age seventeen he entered a government competition for an underwater propelling device. Gatling submitted a design almost identical to the screw propeller that John Ericsson had patented a few weeks earlier.

Undeterred, Gatling invented a successful seed-sowing machine that, when adopted as a wheat drill, made him world famous. He married Jemima Sanders in 1854 and moved to Indianapolis, where he continued to produce new inventions. By far the most important was the multibarrel machine gun that bore his name. Patented in 1862

Watercolor, *The Gatling Guns in Action,* by Charles Johnson Post. (Smithsonian Institution)

and adopted by the U.S. Army in 1866, it remained in American military service until 1911.

The French mitrailleuse (machine gun) reversed the method: Cranking successively discharged the cartridges in twenty-five fixed barrels encased by a wrought iron frame. Its salient shortcoming was comparatively great weight. Because mitrailleuses required horse-drawn carriages, in fact, the French army regarded them as artillery pieces. Consequently misused as light guns rather than in close support of infantry, they forfeited their potential impact during the Franco-Prussian War by being employed at excessive ranges.

The first true machine gun—that is, a weapon automatically self-actuated rather than externally powered—resulted from the work of Hiram Maxim, an American expatriate in London. Between 1883 and 1885 he patented virtually every imaginable method for automatic fire. His basic patent, dating from 1884, covered a recoil-operated lock breech system applied to machine guns. The formerly wasted power of the gun's recoil loaded, fired, and ejected belt-fed cartridges continuously, as long as the trigger was depressed.

So well conceived was Maxim's basic design that it remained virtually unchanged throughout its long career. At its inaugural demonstration in 1885, Maxim's first model fired over 600 rounds per minute through its single barrel. Enclosing the barrel in a water-cooling jacket solved the problem of dispersing the excessive heat generated by firing so many bullets so rapidly. Relatively lightweight because it dispensed with multiple barrels, it also used a tripod rather than the wheels of multiple-barrel systems. Machine guns became much harder to confuse with field artillery.

By 1897 several armies had adopted Maxim guns, but doctrine for their effective combat use lagged behind the proliferation of automatic weapons after Maxim's breakthrough. Although the machine gun was essentially an American invention and the models most widely used in World War I and after all bore American names—Maxim, Hotchkiss, Lewis, Browning— U.S. forces clung to Gatling guns until the turn of the century. Only in 1904 did the U.S. Army adopt Maxim guns, and not until the World War did it acquire machine guns in any great numbers.

QUICK-FIRING GUNS

The same advances in fixed ammunition, smokeless powder, and metal-lurgy that revolutionized small arms also transformed artillery. Quick-firing guns first appeared after the Civil War in light artillery pieces modeled on the Gatling principle, as exemplified by the revolving cannon invented by Benjamin Hotchkiss, adopted by the armies of several countries, including the United States. Hotchkiss began with explosive shells, which became a mainstay of the Union Army in the Civil War. After the war, Hotchkiss moved to France to produce weapons and explosives for the French government. The company he founded in 1882 was headquartered in the United States, but had factories in the countries that bought his products: France, England, Germany, Austria, Russia, and Italy.

Meanwhile, inspired by Maxim's machine gun, Hotchkiss and other inventors quickly followed suit. The Hotchkiss heavy machine gun became the mainstay of French armies in both world wars. But they also applied recoil, then gas operation to light artillery pieces. These light, quick-firing guns found a naval niche, as a defense against torpedo boat attack, but armies were less enthusiastic until the Boers showed the world what pom-poms (the nickname they acquired in the Boer War) could do against troops in the open.

For relatively light projectiles, upgraded machine guns firing 1- or 2-inch shells could absorb enough of the recoil to remain reasonably accurate. Controlling recoil from larger shells was the last step in perfecting the field gun: Such control allowed the gun to remain on target and so greatly increase its rate of accurate fire. The French Model 1897 75-mm gun, the storied "Seventy-five," was the first and most successful design. Hydraulic braking absorbed the recoil and compressed air pushed the barrel forward to its original position. Because the carriage remained fixed, laying and firing the gun became a continuous operation, rather than a process to be repeated from the beginning after each shot. Another benefit of the fixed carriage was that it could mount a shield, thus offering the crew some protection against small arms fire.

Nickel-steel barrels good for 6,000 rounds, interrupted-screw breeches that combined easy manipulation with great reliability, brass-cased fixed ammunition, shells filled with high explosive rather than black powder, flat trajectories yielded by smokeless powder, all joined to make the French 75 an incomparable field gun. A well-trained crew could fire as many as thirty well-aimed shots a minute with devastating effect. The same technological advances that contributed to the radical renovation of field artillery also revolutionized big guns. Although they retained their ancient roles in the attack and defense of fixed fortifications, formal siege warfare no longer remained their exclusive activity. Greater mobility, increased rates of fire, and longer range allowed heavy artillery to assume much larger roles in field operations, as the World War would demonstrate.

By the turn of the century, the eclipse of artillery by small arms that had so colored the Civil War was over. Artillery allied with machine guns, not small arms, swept the fields of World War I. In this as in so many other areas of military technology at the turn of the century, U.S. forces lagged behind European. While other European nations promptly followed French example, U.S. guns still fired black powder in the Spanish-American War. Only after the War Department's postwar reorganization did the U.S. Army acquire a near equivalent to the French 75, the Model 1902 3-inch field gun. That U.S. forces adopted the French weapon in World War I owed more to American production shortcomings than to any significant defect in the homegrown gun.

THE PROSPECT OF MOTORIZATION

America's love affair with the automobile trailed Europe's, and the U.S. Army lagged well behind civil society as well as European armies. By the turn of the century, however, individual soldiers were experimenting with vehicles at their own expense—Illinois National Guard Maj. R. P. Davidson, for instance, mounted machine guns on several modified motorcars, both gas- and steam-powered, between 1899 and 1902. Although they worked well enough and aroused some high-level interest, the War Department rejected their use. This was not simply a matter of stubborn conservatism. Early-twentieth-century vehicles were none too reliable, paved roads outside cities were few, and army funds fell short of basic needs, much less experimentation.

Though hampered by lack of funds, the Quartermaster Department in 1911 began working on truck designs for field use. Ultimately, however, commercial trucks, already becoming common in urban areas, proved more than adequate for the purpose. The 1916 Mexican expedition under Gen. John J. Pershing relied heavily on motor vehicles for supply in rugged, largely roadless terrain. Pershing also made substantial use of armored cars. Notwithstanding operations, maintenance, and personnel problems, the experience confirmed the economy and efficiency of motor transport. Lack of funds might slow the replacement of animal-drawn wagons by self-propelled trucks, but the issue was no longer in doubt. Pershing's expedition also flew six reconnaissance planes, the army's first use of heavier-than-air machines under field conditions. Despite the loss of all six within a few weeks, results seemed promising.

In 1908, less than five years after the Wright Brothers proved the possibility of power flight in a heavier-than-air machine, the U.S. Army purchased the world's first military airplane, a Wright Flyer, directly from the brothers. For almost a decade, the U.S. Army Signal Corps retained control of army aviation. Like so many other late-nineteenth- and early-twentieth-century innovations of military significance invented by Americans, the military airplane developed far more rapidly in Europe than in its homeland, especially after 1914 and the accelerated development spurred by wartime exigencies. The U.S. Army's early lag in military aviation was largely a product of congressional parsimony. As with so many other areas experiencing rapid technological change in the decades before the First World War, those responsible for U.S. military aviation watched from the sidelines.

Wilbur Wright (1867–1912) and Orville Wright (1871–1948)

Their almost universal designation as the Wright Brothers does Wilbur and Orville an injustice. No less individuals than their two older brothers and their younger sister, Wilbur was the confident and steady one, Orville impulsive and enthusiastic. But their merged

The Wrights (left to right): Orville, Katharine, and Wilbur during flight demonstrations at Pau, France, 1909. (Rights reserved by Musée de l'Air et de l'Espace)

identity also reflects a certain truth, that together they embodied a singular genius. They began their lifelong business partnership in 1889 as printers, then added bicycle repair, sales, and manufacturing.

When flight piqued their interest in 1896, they began methodically reading everything available. They identified the crucial aerodynamic, propulsion, and control problems and solved them experimentally with kites, gliders, and a self-built wind tunnel. Even after their first powered flight in 1903 at Kitty Hawk, North Carolina, the brothers continued to improve the design of their machines and seek patents. The crucial year was 1908. The U.S. Army's purchase of a Wright Flyer and Wilbur's flight demonstrations in France left no doubt that the brothers had realized the age-old dream of powered flight.

4

The Naval-Technological Revolution and the Rise of Navalism: Late Nineteenth and Early Twentieth Centuries

◆

In the closing decades of the nineteenth century and the early years of the twentieth, the professionalization of the officer corps in both army and navy was well under way, if not yet fully complete. At the same time, increasingly rapid technological change and burgeoning industrial capacity melded with novel ideas of sea power and navalism to produce a naval arms race. The United States once again assumed a place among the leaders of naval development.

During the 1880s and 1890s, the U.S. Navy began rebuilding itself as a modern sea force. Naval revival fostered economic growth, particularly in steel and related industries, and drew heavily on the growing ability of scientifically oriented engineering to produce specified armor and equipment to order. It also contributed to the rise of navalism based on an exciting new theory of sea power. America's imperial venture opened auspiciously with the successful naval war against Spain. Sustained in part by the growing scientific competence of American medicine, United States forces proved capable of fighting and building in the tropics. The Panama Canal symbolized the country's new status in the world.

MILITARIZATION OF INDUSTRY
AND EDUCATION

Professionalization of staff officers in the War and Navy Department's technical and supply bureaus followed a different path than that of line officers. Less commanders than managers, they had charge of the design, development, production, acquisition, distribution, repair, and disposition of everything the armed forces required for their operations. They also directed the service's personnel systems. They managed depots, shipyards, and armories, designed and made ordnance, contracted with commercial and industrial firms for a variety of goods and services, transported troops and supplies, maintained communications, and provided for the treatment of the sick and wounded. Their technical expertise established staff officers as peers of civilian entrepreneurs and academics, but too often separated from the line. As the bureaus consolidated their power in the late nineteenth century, they tended to favor a most cautious approach toward innovation.

The very process of industrialization itself owed no small debt to military interests. A career army officer and arsenal manager, Capt. Henry Metcalfe, wrote the first book on factory management ever published in the United States. He addressed his 1885 work, an acknowledged classic of management, less to fellow officers than to corporate managers. The U.S. Navy's choice of industrial contractors to supply armor and ordnance for its new fleet in the late nineteenth century, rather than rely on government facilities, may well mark the beginning of what has more recently been termed the military-industrial complex. From the early-nineteenth-century uniformity system to the late-twentieth-century U.S. Air Force sponsorship of automated machine tool development, key aspects of industrial technology have emerged from military settings.

But military example may have been even more important in furthering the regimentation of the labor force that industrialization required. Discipline was the key, military practice its inspiration. Military regimentation inspired the factory system and created the conditions for its implementation. Entrepreneurs and captains of industry found much to admire, and to adopt, in the regimentation and redivision of labor imposed on modernizing armies. Frederick W. Taylor reshaped American, and eventually worldwide, industrialism when he devised scientific management in the late nineteenth and early twentieth centuries. Taylor saw his problem as organizing and directing the work of a mass of socially isolated, ill trained, and poorly motivated proletarians. Holding a mechanistic image of human behavior, he found the answer in regimented action. For Taylor, it was a two-way street. Military models influenced his reforms and scientific management, or Taylorism as it has often been termed, found a receptive military audience.

Similar patterns marked America's schools. Military concerns and money, sometimes direct, sometimes funneled through corporate intermediaries, affected higher education in many ways. Military training on campus dated to the Morrill Act of 1862, though it became institutionalized in the form of the Reserve Officer Training Corps (ROTC) only in 1916. Engineers figured prominently among supporters of this contested institution. American universities, like much of American society, for the most part welcomed such values. After he became the first head of the University of Illinois in 1867, John M. Gregory regularly cited the value of military order and drill in higher education: It promoted discipline, built character, and generally improved the tone of the campus.

Youthful Americans would also come to enjoy such virtues, which evolved in secondary, even primary, schools, later bolstered by Junior ROTC and vocational education. To many Americans, that seemed both worthwhile and desirable. Discipline derived from military training in public schools was alleged to instruct pupils in civil government and respect for law. Such lessons held no less value for adults. Militarism could be one side of the coin, but civic virtue and patriotism might be the other. In many respects, a similar pattern prevailed in England, and was even more widespread on the European continent. Great as the influence of such values might have been, however, they appeared to have little direct impact on science and engineering in the universities, at least before World War I.

NAVAL PROFESSIONALIZATION
AND NAVALISM

Established in Annapolis, Maryland, in 1845, the school that became the United States Naval Academy trailed its army counterpart by two generations. The naval academy in fact took West Point as its model. And like the military academy, the naval academy owed its existence, at least in part, to a perceived engineering need. Naval training had traditionally taken the form of apprenticeship at sea, a practice that persisted long after the naval academy's founding. As steam propulsion challenged the navy's traditional reliance on sail, qualified steam engineers assumed ever-greater importance in the navy. In time, the naval academy would become to mechanical engineering what the military academy had been to civil engineering.

Through much of the nineteenth century, however, the tension between line officers and engineering officers remained one of the central features of naval life.

The number of midshipmen increased sharply during the Civil War, but the next two decades are often labeled the navy's dark ages. The United States discarded one of the world's most advanced navies after 1865, seeing no plausible enemy against which so powerful and expensive a force might be needed. The navy returned to sail. Although ships carried auxiliary steam engines, burning coal verged on a court-martial offense. Like the army isolated on the frontier, the navy isolated on distant stations proved fertile ground for growing professionalism. Then came the renaissance of the 1880s and the budding alliance of naval intellectuals, industrialists, and politicians. Congress authorized new steel ships, industrialists received contracts for armor plate, and intellectuals got the Naval War College.

Under the leadership of Stephen B. Luce, the Naval War College quickly established itself as a center of new thinking about every aspect of naval affairs from policy to tactics. From it issued the world-changing theory of sea power, the idea that great nations flourished on seaborne commerce that could only be sustained by strong navies. It was the brainchild of Alfred Thayer Mahan, son of one of West Point's most influential instructors, Dennis Hart Mahan, and an early graduate of the naval academy. A successful twenty-year career brought him to the new Naval War Colleges at Luce's invitation. His course of lectures in naval history, published in 1890 as *The Influence of Sea Power upon History*, provided the ideological underpinnings of navalism and the technological arms race that dominated the decades preceding World War I, infecting even a United States remote from Old World frictions. Mahan wrapped navies, commerce, colonies, and national power into a neat package that excited imaginations throughout the Western world and beyond.

Stephen Bleecker Luce (1827–1917)

Born in Albany, New York, and raised in Washington, D.C., Luce joined the navy in 1841 as a midshipman. After four years at sea, he entered the newly formed U.S. Naval Academy, graduating in 1849. In 1854, shortly after receiving orders for a three-year tour of duty with the Coast Survey, he married his childhood friend Eliza Henley. The couple had three children. In 1860 Luce returned to the naval academy as instructor in seamanship and gunnery. His textbook on *Seamanship* remained standard at the academy for the rest of the century.

During the Civil War, he alternated teaching with sea duty and began promoting more systematic naval training. His numerous articles on this and other subjects made him one of the navy's intellectual leaders in the postwar decades. By the early 1880s he had emerged as a forceful advocate of naval postgraduate education, culminating in the 1884 establishment of the Naval War College at Newport, Rhode Island, with Luce as its first president. He departed in 1886 for other duties, but returned to Newport after his 1889 retirement and remained active in War College affairs until 1910.

THE STEEL NAVY AND THE NAVAL INDUSTRIAL COMPLEX

The increasing professionalization of the naval officer corps and the growing influence of Mahan's theories of sea power came at a time when naval technology was in the throes of revolutionary change. Wooden sailing warships had long since become obsolete as steam replaced sail and iron, wood, the whole encased in armor. As masts and sails vanished, guns moved from the fixed broadside to rotating turret, and breechloading rifles replaced muzzleloading smoothbores. In the last half of the nineteenth century European navies competed in building ships with ever-thicker armor and ever-larger guns.

In contrast to an often-pioneering role in early-nineteenth-century naval innovation, the United States lagged after the Civil War. Only late in the 1880s did the U.S. Navy reenter the naval era it had helped inaugurate in the Civil War. Naval rebuilding began in the 1880s, at first without any very clear direction. When Benjamin Franklin Tracy became Secretary of the Navy in 1889, he used the office to translate Mahan's new ideas about sea power and strong navies into American policy. Besides strongly supporting naval reform, professionalization, and modernization, he personally helped forge links between the government and the steel industry.

Coincidentally, Mahan's *Influence of Sea Power* appeared in 1890, the same year the U.S. Navy launched its first modern warship, an armored cruiser, and received congressional authorization for its first three battleships. When launched in 1893 they became the first American warships to compare with contemporary European designs, fast, heavily armed, and belted with 18-inch armor. Though also of doubtful seaworthiness and range, they acquitted themselves handsomely in their one trial by fire, the Spanish-American War. Scientifically and technically, nothing much of note emerged from America's "splendid little war," the tag applied by U.S. Ambassador to England John Hay in 1898. The war was largely decided at sea by the quick and overwhelming victories of the new steam and steel navy. Ashore, fighting against largely unenthusiastic, if not demoralized, foes, the army also won quick, easy victories, despite appalling organizational and logistics problems.

MEDICINE AND EMPIRE

Despite the success of American arms, the Spanish-American War was, from a medical viewpoint, a near disaster. Overall, the ratio of deaths from disease to battle deaths reached, by some reckonings, 7 to 1, a much worse record than that achieved during the Civil War. In part this may be attributed to the war's brevity. Inexperienced soldiers—the regular army's strength in 1898 was under 30,000, while a quarter of a million men served in the war—have always tended toward high disease rates until they learn better. The war was too short to show any benefits from improving camp hygiene and sanitation.

Typhoid had been the main killer during the Spanish-American War. Convened after the war, an army medical research board asked why. Walter Reed and his colleagues concluded that the disease spread mainly by flies, by contact between persons, and by human carriers. Sanitary measures offered the best hope of controlling the disease at that point, but a decade's research produced an even better answer: Vaccination with killed bacilli largely eliminated the threat of typhoid. Successful tests in 1909 resulted in vaccination becoming compulsory for the entire U.S. Army in 1911.

Walter Reed (1851–1902)

Raised in Virginia, Reed attended that state's university briefly as an undergraduate before entering its medical school, from which he graduated in July 1869, not yet having celebrated his eighteenth birthday. He earned a second M.D. at Bellevue Hospital Medical College, New York, the following year. In 1876, a year after receiving his Army Medical Corps commission, Reed married Emilie Lawrence, with whom he had two children.

Reed embarked on his medical career just as the germ theory of disease was revolutionizing the field. In 1893 he went to Washington as curator of the Army Medical Museum and professor at the new

Print by Dean Cornwell, *Conquerors of Yellow Fever*, shows Dr. Walter Reed, center, in Cuba. (USPHS, National Library of Medicine)

Army Medical School. In 1898 rampant disease in Spanish–American War army camps precipitated a medical crisis. Appointed head of a medical board, Reed determined that typhoid fever spread by flies and unsanitary conditions was the main culprit. As head of another board, he then identified mosquitoes as the carriers of yellow fever, which threatened American occupation forces in Cuba. These discoveries laid the groundwork for the practical measures that virtually eradicated these diseases.

The Spanish–American War also confronted the army with tropical diseases for which it was ill prepared. After the war, the army's responsibility for administering the new dependencies—and, in the Philippines, suppressing what it labeled an insurrection—brought increased urgency to solving the puzzles of tropical disease. Army medical research boards again provided answers. The best known was Walter Reed's Yellow Fever Commission in Cuba, which by early 1901 proved the disease's cause to be a mosquito-borne virus. British research in the 1890s had shown that mosquitoes also transmitted malaria-causing microorganisms.

These findings did not reflect any profound breakthrough in basic knowledge; much remained to be learned about the etiology of both diseases. What the Army Medical Department accomplished remains nonetheless important: The army acquired practical means of coping with the two main tropical diseases, and a number of lesser ones as well. Anti-mosquito measures quickly became the basis for controlling and preventing both malaria and yellow fever. Acting on the Reed commission findings, William Crawford Gorgas succeeded in reducing malaria and eliminating yellow fever in Havana by the end of 1901.

Major Gorgas then turned to a survey of the proposed site in Panama for a canal linking the Atlantic and Pacific. Endemic malaria and yellow fever jeopardized that plan, he warned. When William Goethals took charge of the project in 1904, he appointed Gorgas as chief sanitary officer of the Panama Canal Zone. Gorgas again succeeded. In two years malaria had been sharply reduced and yellow fever eradicated in the zone. His achievement freed Goethals and his team of fellow West Pointers to concentrate on the task at hand, which they did to magnificent effect. Construction on so vast a scale required a further ten years to complete, but the first ship passed through the Panama Canal in August 1914. It was perhaps fittingly ironic that world acclaim for this triumph of civil engineering aided by science should be muted by the outbreak of war in Europe earlier that month.

BATTLESHIPS AND THE NAVAL ARMS RACE

To a degree quite unmatched in earlier periods, armament making became an entrepreneurial activity in the later nineteenth century. Western armed forces increasingly relied on industrial research for technical innovation, development, and production. In the United States a new naval-industrial complex provided research as well as hardware. American entrepreneurs had never been shy about seeking military contracts, but seldom enjoyed much luck except during wartime. As the earlier experience of Whitney, Hall, and North suggests, the armed forces preferred to rely on their own resources.

The army's arsenal system survived largely intact through the early twentieth century, but the navy forged an alliance with the steel industry toward the end of the nineteenth century. When Congress in 1883 approved building the navy's first steel ships, and in 1886 decreed that only domestic material might be used, it created the first really large and lucrative market for peacetime military industry. The all-steel navy program of the 1880s and 1890s marked a fateful step along the way to transforming the relationship between armed forces and industry.

As the twentieth century opened, the navies of the world were undergoing another, even more far-reaching transformation than had marked the later nineteenth century. The U.S. Navy amalgamated its engineering and staff corps with line officers, ending the long tradition of separate career paths for thinkers, managers, and fighters. One of the beneficiaries of this change was David Taylor, the brilliant naval architect who directed the navy's Experimental Model Basin in the Washington Navy Yard from 1899 to 1914 and whose work did much to complete the conversion of naval hull design from art to science.

David Watson Taylor (1864–1940)

Raised and home-schooled in a Virginia farm family, Taylor graduated first from Randolph-Macon College in 1881, then from the naval academy in 1885. A three-year assignment to advanced study of naval architecture at the Royal Naval College in Greenwich, England, followed. At both Annapolis and Greenwich, Taylor set records for high academic achievement. He spent the next decade in naval construction duties on both coasts. His marriage in 1892 to Imogene Maury Morris produced four children.

Experimental Model Basin building on the waterfront at the Washington Navy Yard, 1901. (U.S. Navy)

In 1899 Taylor became director of the Experimental Model Basin, a facility that used towed models to provide quantitative measures of water resistance. Conducted over a period of fifteen years, his experiments produced a mass of hydrodynamic data that greatly increased the efficiency of preliminary hull designs and enabled naval architects to estimate with considerable accuracy how powerful a ship's engines had to be to reach the speed it was designed for. In 1914 Taylor rose to rear admiral and became chief of the navy's Bureau of Construction and Repair, a post he held until he retired eight years later.

The Royal Navy soon provided perhaps the most potent symbol of the new navalism with the 1906 launching of HMS *Dreadnought*, leading to an even more intense naval armaments race. Enormously fast and heavily armed, battleships and battle cruisers epitomized early-twentieth-century military might. "Dreadnought," like battleship, became a common noun, the design promptly emulated and improved throughout the industrialized world. Two American all–big gun battleships had actually received Congressional authorization in early 1905, before *Dreadnought* was laid down, but *South Carolina* and *Michigan* took four years to complete, in contrast to *Dreadnought*'s extraordinary construction in little over a year. That left *Delaware* and *North Dakota* of 1909 to be generally regarded as America's first dreadnoughts. *Delaware*, the first completed, displaced twice the tonnage of its largest predecessor. She was also 5 knots faster, far longer ranged, and carried more and bigger guns.

TORPEDOES AND SUBMARINES

Impressive as such figures might be, dreadnoughts were also the end of the line for capital ships whose power resided in their guns. The first challenge to their supremacy already existed in the form of torpedoes. Torpedo and mine, all but synonymous terms for stationary or free-floating explosive devices through much of the nineteenth century, achieved their first substantial success in the American Civil War. But only when such devices acquired the means of propelling themselves through the water did they emerge as potentially decisive factors in naval warfare.

Developing the ideas of the Austrian naval officer Johann Luppis, his partner in a Fiume factory, expatriate British engineer and factory manager Robert Whitehead devised the modern self-propelled torpedo in the 1860s. Unimpressive in its early trials—erratic, inaccurate, speed well under 10 knots, range of only a few hundred yards—it nonetheless promised much. Hydrostatic depth regulation proved the crucial innovation. The so-called Whitehead torpedo incorporated an ingenious mechanism that allowed it to maintain constant depth. That meant it could run deep enough to avoid being deflected by surface waves, yet not too deep to pass under its target.

Speed and range improved steadily, as did accuracy when gyroscopic rudder control (an American invention) appeared in the mid-1880s. Not only did Whitehead torpedoes attain 30-knot speeds and 500-yard effective ranges by the turn of the century, but the technology clearly had not reached its limits even then. Initially, the promise of self-propelled torpedoes stimulated the development of torpedo boats, then of torpedo-boat destroyers that usurped their function. But self-propelled torpedoes also provided the first really effective armament for submarine boats.

The turn of the century saw the introduction of the first practical submarines. Combined with the self-propelled torpedo, a centuries-old dream seemed on the verge of realization. Submersible boats, like torpedoes, had some effect in the Civil War. But again like the torpedo, the modern version of the submarine waited until later in the century. The last third of the nineteenth century saw much work on such vessels, especially in France. When the modern submarine appeared at the very end of the century, however, it was largely the work of Irish-American inventor John P. Holland, with a strong assist from his long-time competitor, Simon Lake.

John Philip Holland (1840–1915)

Born and reared in a colonized and poverty-stricken Ireland, Holland hated the British overlords that he saw as the authors of such distress. Aged twelve when his father died, he entered the monastery school in Limerick and excelled in science. Five years later, in 1858, he joined a Catholic teaching order, received further training, and spent the next fifteen years teaching any number of subjects throughout Ireland. But his family had meanwhile moved to Boston and, in 1873, he decided to join them, at least partly because he wanted to build a submarine.

Holland had been sketching submarine designs ever since the American Civil War aroused his interest. Submersible boats might help win Irish independence by countering the Royal Navy. That prospect brought his initial funding from the Fenians, an Irish-American revolutionary group, but other sponsors followed. The five boats he

John Holland and his submarine *Holland*. (U.S. Navy)

built from 1875 to 1897, though flawed, showed real promise. With the sixth, named *Holland*, the promise was realized. Purchased by the U.S. Navy in 1900, it became the first of many to join fleets throughout the world.

Holland won the U.S. Navy's first submarine contract in 1895, but the navy insisted on steam for surface propulsion and the vessel became a much-delayed failure. Undaunted, Holland built his own design, using internal combustion for surface travel, electric batteries for submerged power. The navy liked the result and the vessel bearing his name entered service in 1900. *Holland*'s dual propulsion system featured a 120-horsepower gasoline engine giving her a surface speed of 8 knots and a range of 1,500 miles; with storage batteries driving a 150-horsepower electric motor, she could make the same speed underwater for 50 miles. Horizontal rudder and ballast tanks controlled diving, submerged travel, and surfacing. *Holland* carried two reloads for her bow torpedo tube, and two fixed pneumatic guns intended to hurl so-called "aerial torpedoes," or dynamite shells. Diminutive by later standards, or even compared to most of her contemporaries—10 feet across, 53 feet long, displacing 75 tons—*Holland* was nonetheless the true prototype of the modern submarine.

Like the torpedoes that became their main weapons, the first submarines left plenty of room for technical improvement. As had been true of many other military-technological innovations over the preceding century, the United States soon found itself outclassed as further development of the basic design moved more quickly abroad. However much enlarged and refined over the next half century, though, they retained the basic features Holland gave them, at least until the advent of nuclear-powered submarines in the 1950s.

5

The Catastrophe of Industrial War: 1914–1918

◆

The World War of 1914–1918 confirmed trends clearly evident in retrospect since the mid-nineteenth century. Military institutions had changed dramatically. Repeating rifles, smokeless powder, quick-firing long-range field artillery, and machine guns multiplied firepower and extended the killing zone. Doffing gaudy color in favor of field gray or khaki, soldiers left firing lines and maneuver for ground cover and trenches. Runners began giving way to telegraph and wireless, muscle to steam and petrol. Staffs burgeoned to direct vast armies as nations prepared to put millions of men under arms. With the new giving way to the newer ever more quickly, almost every aspect of military life was altered if not transformed. Equally dizzying changes marked naval technology. So rapid did the pace of change become toward century's end that the ships of one decade seemed almost worthless in the next.

Innovations so radical scarcely passed unnoticed. Military and naval novelties figured prominently, for instance, in popular turn-of-the-century compendia on the progress of invention. Yet judging their likely impact surpassed most contemporary imaginations, military and civilian alike. Indeed, many have blamed the catastrophe of World War I on European armies blind to the meaning of swiftly changing technology. That may be unfair. Innovations in military technology still mostly came from nonmilitary sources. That made the flood of new or improved arms hard to control or direct. Military planners did not so much ignore problems as misjudge their magnitude. Whatever the reasons, the result was catastrophe almost beyond comprehension.

STALEMATE ON THE GROUND

World attention focused on the deadlock that quickly emerged on the Western Front, where the United States would later intervene. Battle ceased being a decisive meeting of moving forces. Mechanized firepower drove troops to ground. Maneuver vanished in trenches and operations took on the guise of a gigantic siege, or a network of interrelated sieges. Like all sieges, war on the Western front locked the foes into close contact and highlighted the work of gunners and engineers. Strategy for much of the war seemed to mean little more than augmenting gunfire; tactics centered on the attack and defense of ever more elaborate field works. Huge armies, in essence, fought continuously. The so-called battles along trench lines from Swiss frontier to North Sea were merely more intense episodes in largely uninterrupted combat. Several thousand casualties accrued every day even when neither side was mounting a major offensive.

Firepower created the deadlock, but the failure of other technologies frustrated attempts to break the impasse. Transportation and communication shortcomings may have been the most critical. Bombardment could and regularly did open the way for infantry to seize forward enemy positions. But when troops advanced they left behind easy links to artillery support and reinforcements. Railroads terminated behind the lines; motorized transport carried some material forward, but men and mules provided most carriage. Beyond the trenches were only the troops themselves to manhandle supplies across broken and trackless ground. Reinforcements likewise had to make their way on foot, exhausting much of their strength before ever reaching the enemy.

Timing often became a major problem, how to obtain artillery support, supplies, or fresh troops when and where needed. There was simply no easy, reliable way to maintain contact across no-man's-land. Telegraph or telephone depended on wires too easily broken when strung at ground level behind moving forces; for fixed positions, of course, the wire could be buried, and even then was often disrupted by artillery barrage. Portable wireless gear unavailable until the final year of the war left only ancient and inadequate techniques: Visual signals were almost worthless, runners or animal couriers too slow and uncertain. Artillery support thus became problematic, as did calls for reinforcement. Defenders falling back on their lines of communication could almost always react more quickly than attackers leaving theirs behind.

As befitted a scientific-industrial age, efforts to break the deadlock called on technology. Technical fixes, however, enjoyed only limited success. Stalemate derived in part from still changing military technology, but even more from the vastly improved artillery and machine guns developed in the decade or two before war began, and perhaps most of all to the enormous productive capacity that made it possible to equip, maintain, and restore huge armies. Yet the apparent role of novel technology in creating the stalemate of World War I suggested to many the promise of searching for technological solutions to the stalemate.

Because several of its most prominent and promptly exploited innovations were chemical—notably toxic gases, artificial nitrates, and mass-produced explosives—World War I has sometimes been called a chemist's war. Among these novelties only gas directly affected the battlefield. Initial success notwithstanding, however, it yielded no decision. Though trenches offered scant protection against gas, masks and better training largely controlled the danger. Extensive use during the war meant no more than the briefest flurry of postwar interest in the future of gas warfare. The army simply reabsorbed the independent U.S. Army Chemical Warfare Service that William L. Siebert, one-time hero of the Panama Canal project and failed division leader in France, had organized in 1918 and commanded until his retirement in 1920.

TRUCKS, TANKS, AND AIRPLANES

It was quite otherwise with a range of mechanical innovations based on the internal combustion engine, a late-nineteenth-century product that laid the groundwork for large-scale changes in the conduct of war. Mechanics, not chemists, had the future in hand. Development along three lines significantly, if not decisively, affected the course of World War I, though holding out still greater promise of future use: motorized transport, armored vehicles, and armed aircraft.

By 1910 the United States had become the world's leading automobile maker. Despite such resources to draw on, American armed forces were slow to recognize motorization's potential. Several European nations had substantial motorcar industries, however, and their armed forces had taken steps toward turning such machines to military advantage. Motor transport, especially, began to influence logistics significantly, though it hardly displaced animal-drawn transport.

When war began, trucks were rare; the United States registered scarcely 10,000 in 1910, at the same time car registrations neared half a million. Allied war orders flooding American industry spurred massive growth.

Financed by government money, American truck production leaped from just under 25,000 in 1914 to well over 200,000 in 1918. Industry produced trucks in permanently higher numbers after the war, with important consequences for the American economy. Standardization and reliability of engines, transmissions, and body designs tested in the harsh school of the Western Front soared.

Motorized transport, much as it contributed to supplying the front lines, could not solve the battlefield stalemate. Various experiments with armored cars proved equally marginal. Wheeled vehicles could not traverse the shell-torn ground that impeded all movement. The answer appeared in the form of the tank, the British code name that track-laying, gun-carrying, armored vehicles have ever since retained. Tanks allowed self-propelled guns to keep up with advancing infantry; tracks took them across broken ground and armor protected them from machine guns. Conceived as siege machines, a technological fix for the problem of breaking through defenses too strong for men alone, tanks promised more than they could yet deliver. Relatively slow, thin-skinned, short-ranged, lightly armed, and fragile, the machines of the Great War could not always achieve even their intended purpose, limited though it was.

Yet they evoked visions of a different kind of future war. Though aborted by the Armistice, the Allied Plan 1919 called for an attack by massed tanks of new design, able to speed to the enemy's rear and paralyze the defense, while fleets of airplanes carried the assault from above. The U.S. Army formed its own independent Tank Corps soon after entering the war. Organized and led by Samuel Rockenbach, a cavalry officer who had served with Pershing in Mexico and France, it proved its mettle in the battles of 1918. Like the Chemical Warfare Service, though, its independence did not long survive the Armistice.

Samuel Dickerson Rockenbach (1869–1952)

Born in Lynchburg, Virginia, Rockenbach grew up with romantic stories of his father's service in Lee's Army of Northern Virginia. He graduated from Virginia Military Institute in 1885. Six years later he won a commission in the U.S. Cavalry. During his career as a junior officer, mostly in the West, he displayed an unusual talent for administration, which often led him to staff, especially quartermaster, duties. More important, he served several times with John J. Pershing.

When Pershing commanded the Mexican punitive expedition in 1916, he requested Rockenbach as his quartermaster officer. And when

Tanks Attacking Early September 26th, charcoal drawing by George Harding, 1918.
(Smithsonian Institution)

Pershing took command of the American Expeditionary Force, he once again sought Rockenbach's services, first as quartermaster, then to head the brand-new Tank Corps, which Rockenbach successfully organized, trained, and led into battle. He remained Tank Corps chief until 1920, when Congress made it part of the infantry. He continued to direct the Tank School at Fort Meade, Maryland, until 1924, when he was promoted to general. Although his military career lasted another decade, he had no further role in tank development.

Flying machines even more than tanks evoked visions of a different kind of war. The airplane was scarcely a decade old when the war began, but rapid progress had followed the success of the Wright Brothers. Reconnaissance from the air quickly became a vital aspect of war-making. Air combat promptly followed, and a technological race for air superiority brought still more rapid improvement. This aerial arms race most clearly portended things to come. Constantly improving aircraft and weapons periodically shifted technological advantage from one side to the other and back again, with direct effects on the war in progress. Technological innovation is not alone decisive; new weapons require new doctrine, tactics, and organization. This has always been true, but in the modern era the pace of change has so increased that the time available has decreased from centuries and decades to years and months.

Airplanes went to war in 1914 as general-purpose flying machines without weapons, able to fly no faster than 70 miles per hour, to remain aloft no longer than an hour. By 1918 both top speed and endurance for the single-place fighter had doubled, and it now sported twin machine guns synchronized to fire through its propeller. And fighters had become but one of several specialized types of aircraft, which included ground attack machines, light bombers, and huge multi-engine long-range heavy bombers like Germany's Gotha.

Tactically, attacks on ground forces from the air produced only modest results during the war. Aerial bombing of factories and civilians, in contrast, showed real promise. Despite limited numbers of machines, German attacks on London, first with Zeppelins and then with Gotha heavy bombers, spread panic and disrupted the city's work. But raids at intervals of many days or weeks by twenty bombers, each carrying but half a ton of bombs, could not achieve decisive results. Notwithstanding such limited results in practice, the potential of such attacks using far larger numbers of much more powerful aircraft seemed clear to many, stimulating visions of a war-winning weapon. Plan 1919 included massive long-range bombing attacks on German cities as well as mass tank attack.

WAR AT SEA

The primary exemplars of the prewar arms race, the great battleship fleets of England and Germany, achieved little at sea. Powerful though they might be, dreadnoughts proved all too vulnerable to mines and torpedoes. While one may argue that the mere existence of the British fleet sufficed to maintain command of the sea, and thus the blockade against German commerce, the fact also remains that the ships themselves could rarely venture from their protected anchorages.

Submarines suffered no such restriction. Though little regarded in prewar thought, they achieved notable success against other warships by launching torpedoes against them while submerged. Like so many products of technology geared to military purpose, the submarine in its early years relied more on promise than accomplishment. During the decade and a half before World War I, submarines acquired diesel engines, periscopes, and wireless gear. They also grew bigger, faster, and more seaworthy. Improvement never stopped, but submersible warships had attained eminently practical form just before the outbreak of world war provided a stage to display their newfound prowess.

Even the largest submarine could carry few weapons as large as torpedoes, and none could remain long underwater. The submarines of World War I, indeed of World War II, were not true underwater vessels. Properly speaking, they were diving boats, surfaced most of the time but able to hide beneath the waves for a while. Far slower underwater than afloat, submarines submerged could only lie in wait rather than actively seek their prey. Against surface warships far more heavily armed and armored than any submarine could be, they had little choice; stealth provided their chief defense. Not so against their preferred targets, merchant shipping.

Unarmed merchant ships could be hunted on the surface and attacked with quick-firing deck guns; valuable torpedoes need not be expended on such targets. Submarines did best raiding commerce, not attacking other warships. A handful of German U-boats came close to applying the kind of blockade against England that the huge Royal Navy imposed on Germany. Technology provided no ready answer to this threat, though organization did. Undersea warfare was thwarted by the convoy system. When the United States entered the war, the head of American naval forces in Europe, Adm. William Sims, made antisubmarine warfare a major focus of planning. Despite their quite limited capabilities and ultimate defeat, however, submarines had proved by far the most effective warships of World War I, and the technology of undersea warfare would only improve.

William Sowden Sims (1858–1936)

Born in Canada of American parents, Sims grew up in Pennsylvania. He graduated from the U.S. Naval Academy at Annapolis, Maryland, in 1880. Although he began his career just as America's new steel navy was emerging, bureaucracy and conservatism still ruled. Sims stamped himself as a progressive when he became an outspoken advocate of new continuous-aim gunnery methods in the late 1890s. Appointed Inspector of Target Practice in 1902, he was able to train many officers and gun crews in the new fire-control system.

When America entered the World War, Sims became the head of American naval forces in Europe, a desk job of vital importance. Germany had resumed unrestricted submarine warfare. Sims recognized that much of the British fleet was committed to maintaining a distant blockade of Germany, making American assistance in antisubmarine warfare, especially the convoy system, crucial. Largely due to Sims'

William Sowden Sims. (University of Iowa Libraries)

efforts, the United States diverted major resources to destroyers and other ships that would bolster the war against German U-boats, and placed much of its Atlantic naval force under British command.

THE MEANING OF INDUSTRIALIZED WAR

World War I became the first great industrial war. Manufacturing and logistics mattered more than all other aspects of war-making. Allied and Central Powers alike reorganized themselves to manufacture death. The new weapons demanded ammunition in staggering quantities, but that was only the first crisis. Armies so huge required supplies of every kind on a formerly unimagined scale. "War economy" and "home front" entered the lexicon for the conversion of industrial capacity, for the reorientation of civic life, for the concentration of all resources toward fighting total war. Managing the armies and keeping them supplied exhausted military art and science. Industrial engineering displaced generalship, and attrition became the recipe for victory.

Ultimately, the war was decided by supply. The great battleships that had captured so many minds and consumed so much money before the war rarely ventured to sea, confined by fear of mines and torpedoes. The most significant naval action of the war involved submarine attacks on shipping. Boats able to travel beneath the surface and attack while remaining submerged were an old idea. Like breechloading rifles and machine guns, they became practical only with late-nineteenth-century technology. Despite their relatively modest capabilities by later standards, German submarines threatened on more than one occasion to deprive Britain of its maritime commerce and cut its economic lifeline, especially imported food supplies. Antisubmarine tactics and the convoy system defeated the threat, but it was a close call.

Although more self-sufficient than Britain, Germany had to import certain crucial raw materials as well as significant amounts of foodstuffs. It was thus a country vulnerable to blockade. Merely by existing, the Royal Navy maintained command of the sea. Denied access to key imports, Germany improvised, sometimes brilliantly. The creation of an artificial nitrate industry to replace lost overseas sources was one of the most remarkable scientific achievements of the war. But food could not be improvised. Britain faced the same dilemma, but succeeded in breaking the submarine blockade. It was a close-run race, decided by the exhaustion of the weaker side.

Throughout the war, Germany tended to retain an edge in tactics and operations, the allies in management and logistics. British war production soared; perhaps more surprisingly, so did French, despite the German occupation of France's major coal and iron region. The collapse of Russia in 1917 and its withdrawal from the war held the long-term promise of giving Germany a new breadbasket. Immediately more important, it augmented Germany's strong suit by freeing scores of divisions for redeployment to the Western Front. In spring 1918 an enlarged German army using novel infantry and artillery tactics came close to victory. Its failure owed more than a little to the intervention of fresh American troops in large numbers, many of them transported to Europe in British ships, most of them equipped from Allied, especially French, factories and depots.

Americans had observed events in Europe with mixed feelings. Ambivalent though they may have been about deciding to intervene, there was nothing reluctant about implementing the decision. Much of the pre-1917 debate on mobilization had centered on manpower, and in the final analysis American manpower was the decisive American contribution. From April 1917, when the United States declared war on Germany, until the November 1918 Armistice, the U.S. Army grew from 200,000 to 3.5 million men, 2 million of whom had reached France.

Ambitious plans to arm and equip the American Expeditionary Force from American sources proved less successful. Economic mobilization for World War I demanded an effort far beyond anything ever before attempted in the United States. It began under auspices of the Council of National Defense, established by statute in August 1916 to coordinate industry and resources; coordination became direction as the council mutated into the War Industries Board, which came to control large parts of the economy. Although better organized than such efforts in past American wars, it took too much time. American industry was still gearing up when the war ended, and American forces fought chiefly with French ordnance and flew French and British aircraft.

MOBILIZING AMERICAN SCIENCE

Mobilizing science for the war effort enjoyed equally modest success. As early as 1915 the navy created a Naval Consulting Board to screen proffered inventions for value and practicality. All its members were working engineers and inventors, including the most famous of them all, Thomas Alva Edison, as chairman. Individual inventors, however, proved to have no better grasp of the scientific-technical needs of modern warfare than did most officers; vanishingly few of the suggested ideas even merited discussion, and none reached the battlefield.

The National Advisory Committee on Aeronautics (NACA) built on a more solid base, but still failed to contribute much to the war effort. Established by Congress in 1915, NACA comprised twelve appointed members, including industry and service representatives, who served without pay. Besides advising on aeronautical matters, NACA soon acquired a larger purpose than its name implied. It sought to coordinate aeronautical work among the several government departments, and also directed an aeronautical engineering research laboratory. The Langley Aeronautical Laboratory descended from turn-of-the-century experimental work conducted at the Smithsonian Institution in Washington, D.C. Though already a going concern on a small scale at the Smithsonian, the laboratory needed room to expand. A move to Langley, Virginia, and the building of facilities were still under way when the war ended.

Far more broadly based than NACA or the Naval Consulting Board, a revitalized National Academy of Sciences seemed the most likely choice to centralize war research. The academy was remembered, somewhat inaccurately, as having successfully performed such a function during the Civil War. Although the academy's true role had been quite limited, that memory nonetheless inspired action. In 1916 the academy formed a National Research Council (NRC) with members drawn from governmental, academic, and industrial research. The council sought to promote long-term basic research as well as work on projects immediately relevant to the war. One such project was submarine detection, which the scientists of the NRC pursued far more successfully than the supposedly practical inventors of the Naval Consulting Board, chiefly due to the energetic and effective Robert A. Millikan.

Robert Andrews Millikan (1868–1953)

Millikan left Illinois, where he had been born and raised, in 1886 to study physics at Oberlin College in Ohio. He completed his doctorate at Columbia University, New York, in 1895, followed by a year of postdoctoral study in Germany. He joined the University of Chicago faculty and in 1907 began the research that culminated in 1913 with the precise determinations of the charge of the electron, and in 1916, of Planck's constant. These two results earned him the 1923 Nobel Prize in Physics.

On the eve of American entry into World War I, Millikan agreed to serve on the National Research Council, becoming chair of its committee on submarine-detection devices. Determined to make science relevant to national preparedness, he set up a research-and-development facility in Connecticut and induced ten of the country's most eminent physicists to join the effort. Although convoys mainly solved the submarine problem, Millikan's team devised and saw

Robert Millikan with the Millikan-Neher self-recording ionization chamber, an electroscope designed to measure cosmic ray intensity. (Courtesy of the Archives, California Institute of Technology)

deployed the war's best, if still inadequate, detector by the summer of 1918. Even so limited a success convinced several high-ranking officials that science would play a major part in any future war effort.

Early in 1917 the Council of National Defense made NRC its department of research, while assigning the Naval Consulting Board exclusively to evaluating inventions. Mechanisms for shifting military funds to civilian research had not yet been devised, however, and the need for haste seemed to preclude delay. Accordingly, junior scientists were conscripted to work on specific projects, leaving little scope for balanced program or detached judgment. Scientific mobilization, like economic mobilization, went more smoothly than it had in the past; it was likewise only beginning to achieve results when the war ended.

What mattered perhaps more than the relatively limited wartime accomplishments was the promise of greater things to come. While the inventors and practical men of the Naval Consulting Board clearly represented a bygone era, scientists like Millikan demonstrated that their apparently esoteric knowledge could produce practical results. The First World War marked the beginning of a new relationship between scientists, who had their first taste of relevance in the corridors of power, and military officers, some of whom recognized that the centuries-old promise of science to revolutionize war might now be nearing realization. Both scientists and officers would prove ready to take up the challenge when the next crisis loomed.

MEDICAL ASPECTS OF TOTAL WAR

Medical services, like other aspects of American military organization, displayed strikingly greater efficiency in 1917 and 1918 than they had in earlier American wars. Reorganization after the Spanish-American War contributed to the improvement, including the establishment in 1901 of the Army Nurse Corps, and in 1908 of the Navy Nurse Corps. America's long-anticipated decision to enter the war also offered time to prepare. During this period of grace, the Surgeon General's Office worked with American Red Cross to form army-model base hospital units, their staffs drawn from the nation's major hospitals and medical schools. Catherine Stimson, who was to become superintendent of the Army Nurse Corps, entered military service as chief nurse of Washington University Medical School's Base Hospital 21. Its staff was drawn from two teaching hospitals in St. Louis, Missouri.

Julia Catherine Stimson (1881–1948)

Stimson was born in Massachusetts, but grew up in New York City. Her father and mother both came from old New England families with

Julia Catherine Stimson. (Library of Congress)

strong traditions of public service and high expectations for their seven children. All four daughters went to Vassar and three earned graduate degrees. Stimson began graduate studies in zoology at Columbia University, but shifted to nursing. After graduating from the New York Hospital Training School for Nurses in 1908, she pioneered medical social work, first at Harlem Hospital, then as superintendent of nurses at two teaching hospitals in St. Louis, Missouri. She also found time to take a master's degree in sociology.

A member of the Red Cross since 1909, Stimson in 1917 became chief nurse of Base Hospital 21 in France, which the Red Cross had organized with Washington University Medical School for the army. She rose to chief nurse of the American Red Cross in France, then director of the American Expeditionary Force's Nursing Service. She returned from France in 1919 to become dean of the Army School of Nursing and acting superintendent of Army Nurse Corps, a position that became permanent the following year and which she held until 1937.

The first steps were also taken to build a medical-instrument industry almost from scratch to replace lost European sources. Like all other elements of the armed forces, the Medical Corps expanded enormously in numbers and complexity during the war. It was aided by a host of civilian volunteers in uniform, many of them women. Tens of thousands of uniformed civilian women and men supported the troops and worked in a wide range of relief and welfare organizations throughout Europe. They drove ambulances, aided the rehabilitation of the wounded, and performed a host of other vital services that contributed significantly to the striking improvement in military health during the First World War.

Improved medical practice and more efficient organization continued to reduce the number of troops who died of their wounds. For the Union army in the Civil War, the ratio of deaths in battle to death from wounds had been 3 to 2. The ratio improved to 2 to 1 in 1898, and 3 to 1 in the First World War. Medical supply services, indeed supply services of all kinds, also attained a level of performance notably better than in the past. They largely managed to keep up with the demands of rapidly swelling armed forces.

Nothing resembling the epidemics that raced through 1898 training camps afflicted the recruits of World War I. The record remained equally good overseas, despite appalling conditions on the Western Front. Then, in the fall of 1918, everything changed. A new, virulent, and highly contagious strain of influenza dubbed Spanish flu appeared almost simultaneously in troop staging areas on both sides of the Atlantic and spread rapidly. It debilitated the warring armies, but struck civilians as well as those in uniform. Sweeping around the world during 1918 and 1919, it killed as many as 40 million people and shook the confidence of a medical profession that could neither identify the cause of the disease nor offer any effective treatment.

The Medical Corps took responsibility for more than the health of the troops. It also conducted the physical examinations of the millions of volunteers and draftees who entered the armed forces, as well as their mental evaluations. One of the war's most striking innovations was the introduction of a scientific method to sorting those millions of inductees and to assign each of them appropriately to one of the increasingly numerous and varied jobs required in a modern army. With help from the American Psychological Association, the army introduced the first large-scale administration of intelligence and vocational tests. The scale of the military experiment and its results were widely perceived by the public and professionals alike to validate the idea, formerly greeted with skepticism or indifference, that intelligence and skills could be quantitatively measured. After the war, such tests gained a significant and growing role in academe and industry.

6

From War to War: 1919–1941

◆

The armed forces had begun to recognize the possibility of directed innovation—that is, the possibility that weapons or weapon systems could be designed and built not merely to meet military needs but to realize military desires—in the late nineteenth century, but that insight became firmly lodged only after the end of the First World War, and triumphed only after the Second. Institutional promises of technological breakthroughs from directed research resided in a flourishing National Advisory Committee for Aeronautics (NACA) and an expanding Naval Research Laboratory during the interwar period. Both provided concrete evidence of how directed innovation might be organized and its benefits reaped. In sharp contrast to the First World War, the United States entered the Second as a leader in aeronautical design and development. No less important when the United States entered the war, NACA also became the model for organizing American war research.

SCIENCE DEMOBILIZED

The First World War imposed extraordinary demands on the economies and societies of the belligerents. Every state had to reorganize its people and resources to sustain its home front, cope with shortages, maintain production, and keep supplies flowing to the war fronts. Bureaucratic management formerly limited to individual business firms or government agencies now applied to the entire state and displaced free-market capitalism at the center of war economies. Science took its place among the activities the state sought to control, with varying success. In the United States these trends were less marked than in other active belligerents, chiefly because of the country's late entry. Scientific mobilization, like other aspects of the American buildup, began well by the standards of former American wars but followed past patterns in ending promptly with the war. So heavily reliant on scientists in uniform, direct military research shared the fate of the armed forces. Armistice brought abrupt cuts in military appropriations and rapid demobilization.

The Naval Consulting Board expired, but left a legacy. In 1916 it had persuaded Congress that the navy needed a research laboratory and secured the funds to begin building. Internal squabbling over a site, however, as well as other pressing concerns, deflected the effort; the Naval Research Laboratory remained in limbo until 1923. Once established, its research program progressed nicely from small beginnings, most notably during the interwar period in work on radar. The National Research Council survived the war. Although it lost whatever function in military research it may have had, it did acquire the congressional mandate it lacked during the war and maintained significant links with the scientific community. Ultimately, NRC would again assume a larger role in the conduct of American research, but that occurred during and after World War II.

Of all the First World War science agencies, only NACA flourished beyond the war's immediate aftermath. Conducting basic and applied research for military and civilian clients at the Langley laboratory, it pioneered most interwar advances in aeronautical engineering. NACA's success owed much to the efforts of George W. Lewis, its director of aeronautical research from 1924 until his retirement in 1947, who built a program with strong emphasis on innovative research. By the end of the 1930s, NACA research had moved the United States to the forefront of aeronautical design and development. The agency's success would also make it a model for organizing American scientific research when war came again.

George William Lewis (1882–1948)

Lewis grew up and went to school in Scranton, Pennsylvania. In 1908 he and Myrtle Harvey married, a union that produced six children. That same year he graduated from Sibley College of Engineering, Cornell University, with a degree in mechanical engineering,

NACA Aircraft Engine Research Center Hangar at Lewis Research Center, 1945.
(NASA)

which he followed with a master's degree in 1910. He spent the next decade teaching mathematics at Swarthmore and heading a Philadelphia company's engineering department.

In 1919 he joined the recently formed National Advisory Committee on Aeronautics as executive officer, becoming director of aeronautical research five years later. In that position he helped transform American aeronautical research and development. He actively recruited young scientists, engineers, and mathematicians into government research. The pay might be lower, he admitted, but they would work on the forefront of research and development. And so they did. In his quarter century at the helm—ill health forced him to resign in 1947—NACA research pioneered almost every advance in aircraft design, development, and testing.

Meanwhile, though, less direct consequences of scientific mobilization for World War I may have been more significant. Colleges became enthusiastic centers for military training and education. They were scarcely less eager in peacetime. During the First World War era, American schools of engineering used military models in organizing research laboratories, adapted their curricula to meet military demands, and borrowed military test methods to evaluate their students. Consequences might be as subtle as those entailed in using personality tests and job specifications created for military purposes during the war. At the other extreme lay the overt effects of research channeled by military funding. By whatever paths, military values of order, discipline, and hierarchy pervaded engineering and persisted after the war ended. Industry readily followed suit. Military techniques of classifying jobs and sorting personnel required little adjustment to suit the precepts of scientific management.

VISIONS OF AIRPOWER

Between the wars, new doctrines of airpower and military mechanization provided the framework for rethinking operations, tactics, and organization. Accommodating newly available technologies derived from
directed research was only one problem. Advocates of military mechanization and airpower played a key role in furthering the idea that this
relationship could be reversed, that military desires from tactical to strategic
might be achieved with machines and weapons designed and built to promote military visions of future war. They constantly stressed the future
promise of technology over its current manifestations.

Uniquely among the technologies of World War I, aircraft and tanks
inspired long-range thinking. Actual wartime experience served more to
spur reflection than to provide concrete example. Visionaries nonetheless
invoked the power of modern science and technology to create future
bombers vastly more powerful than the frail machines that then existed.
Procured in numbers large enough, modern heavily-armed bombers of
great range and large capacity could win the next war.

The production of munitions had reached a level in the World War
that seemed to render battlefields all but irrelevant to the outcome; factories and workers suddenly emerged as prime military targets. After seizing
command of the air with its aerial fighting forces, a nation might destroy a
foe's factories from the air and so deprive its armed forces of the where-
withal to fight on. Rendered defenseless, the enemy might thus be defeated
on the home front rather than the battlefield. Airpower would simply
bypass the stalemate on the ground, and victory would follow inevitably—
once the appropriate strategic bombing aircraft became available. That was
a long time coming, and the results were not everything the prewar visionaries might have wished.

These ideas, along with the demand for an independent air force their
implementation required, received their first popularization in the writing
of an Italian aviator, Gen. Giulio Douhet. In the United States the vision
of victory through airpower was intimately related to the ongoing bureaucratic and public relations struggle of the army's fliers for an independent air force. One of America's flying heroes of the World War, Gen.
Billy Mitchell, followed Douhet's lead. He scored a public relations coup
with his 1921 demonstration that bombs could sink a battleship. Later
court-martialed for insubordination, he resigned from the army but his
disciples remained in uniform, ultimately to achieve everything Mitchell
had advocated. Advocacy was not, however, limited to men in uniform.

Among the most effective was a Russian émigré, Alexander de Seversky, who became a leading aircraft manufacturer in the United States and a forceful advocate of airpower. His 1942 best-seller, *Victory through Air Power*, became an animated Walt Disney film seen by millions.

Implementing airpower doctrine demanded a new kind of airplane. Not until 1935 did the prototype of a strategic bomber actually fly: the Boeing XB-17, called the Flying Fortress. Uniting defensive armor and machine guns, large bomb capacity, long range, and great speed, it could take advantage of the highly accurate Norden bombsight introduced in 1931 to carry the attack deep into an enemy's homeland. Or so the airpower enthusiasts claimed. In 1939, as the air force accepted deliveries of its first B-17s, a second heavy bomber, the Consolidated B-24 (Liberator), made its maiden flight, with first deliveries the following year. Both bombers would see extensive service in World War II, but fall far short of realizing the hopes of airpower advocates.

ARMORED DOCTRINE

In fact quite different uses of aircraft shaped the war to come: fighters and bombers designed to provide close tactical and operational support for sea and ground forces. Developing the machines for such cooperation formed a central theme in interwar development. Two areas, in particular, deserve attention: the decisive contribution of ground-attack aircraft to the spectacular successes of armored forces in World War II, and the radically new role of aircraft in sea warfare based on the development of aircraft carriers.

Like aircraft, tanks inspired potent visions, though less strategic than tactical or operational. And like the doctrine of strategic bombing, the doctrine of mechanized (or armored) warfare invoked the power of modern science to produce the required machines. "There is nothing too wonderful for science," exclaimed J.F.C. Fuller in a prize-winning 1920 essay. After serving as chief of staff in the British Tank Corps during the war, Fuller became the leading postwar advocate of mechanization. "We of the fighting services must grasp the wand of this magician and compel the future to obey us," he insisted.

The British Army was first to hear the message, conducting large-scale armored training exercises in the late 1920s and early 1930s. Although it then seemed to lose enthusiasm for the new techniques, its example inspired widespread emulation, notably in Germany where mechanized doctrine contributed significantly to rebuilding and reequipping the

Wehrmacht, the German Army. Armored doctrine also aroused resistance. Opponents included officers romantically attached to the horse and others whose motives might be suspect, but they also numbered those who raised well-reasoned objections. Mechanized forces would be costly to provide, hard to maintain, prone to rapid obsolescence, and difficult to supply. Only the experience of war clinched the argument by showing conclusively that such obstacles could be overcome, adequately if not always easily.

From the beginning the vision of mechanized war won converts in the United States as elsewhere. The Tank Corps formed in World War I lasted only until 1920. Vehicles dispersed in small packets among the infantry, to say nothing of sharply limited funds, allowed little scope for tactical or organizational innovation, although an experimental mechanized force on the British model underwent trials in 1928. In the 1930s, much influenced by the ideas of Fuller and Liddell Hart, cavalry officers such as Adna Romanza Chaffee, Jr., took the lead in developing concepts of mechanized warfare and experimenting with mechanized units equipped with so-called "combat cars," tanks by another name because by law tanks belonged exclusively to the infantry.

Adna Romanza Chaffee, Jr. (1884–1941)

Chaffee grew up in a military family, the only son in the family of four children of Anna Frances Rockwell Chaffee and army officer, later general, Adna Romanza Chaffee. He and Ethel Warren Huff married in 1908, two years after he graduated from the U.S. Military Academy. They had one child, a son, named after his father and grandfather. During his first decade in the army, Chaffee served mainly in the cavalry, where he instructed mounted tactics in Cuba, the Philippines, and Kansas. An accomplished horseman, he was a valued member of the American riding team. In France during World War I, he attended the prestigious General Staff College at Langres, staying on for a time as instructor.

During the interwar period, Chaffee championed the development of integrated systems of mechanization that modernized the army and helped lay the foundation for the American Armored Force established in 1940. The new organization erased the traditional division of combat forces as cavalry, infantry, and artillery. All its members would ride to battle—in tanks, armored cars, trucks, and self-propelled artillery vehicles—though infantry would dismount for combat. Chaffee's premature death on the eve of America's entry into World War II left the Armored Force to prove itself in other hands.

Little money meant no production. Research and development focused on prototypes and components, a policy that paid handsome dividends when mass production began in 1940, after the army created the Armored Force with Chaffee as its commander. American tank designs stressed simplicity, reliability, and standardization. None equaled the best German tanks they faced, but they were quick to build and easy to maintain. Available in the thousands, lavishly supplied, and skillfully directed, they proved able to win victory after victory even over armored forces equipped with tanks that individually outmatched them. They also enjoyed the inestimable advantage of superb tactical air support.

AIRCRAFT AT SEA

Aircraft also decisively reshaped war at sea. Although the Royal Navy pioneered aircraft carriers, the new kind of warship reached its fullest development in the navies of Japan and the United States. The vast stretches of the Pacific Ocean confronted both nations with operational and logistics problems to which seaborne aircraft seemed at least part of the solution. Although both nations devoted the larger share of their naval appropriations during the interwar years to traditional battleships, both also made real progress on aircraft carriers. In 1921 the navy created a Bureau of Aeronautics and named as its chief William A. Moffett, already a well-known advocate of naval aviation.

William Adger Moffett (1869–1933)

The seventh of nine children in a distinguished family from Charleston, South Carolina, Moffett flourished despite the early loss of his father. For almost three decades after his 1890 graduation from the U.S. Naval Academy, he gained in rank and experience through military assignments and diplomatic missions. In 1902, on shore leave in Southampton, England, he and Jeanette Beverly Whitten of Kingston, Canada, married. They had six children. Moffett received the Medal of Honor in 1914 for service commanding a cruiser during the American invasion of Veracruz, Mexico.

Navy Airship USS *Akron*, Panama, 1933. (U.S. Navy)

While overseeing expansion of the Great Lakes Naval Training Station, Illinois, 1914–1918, Moffett encouraged flight instruction and developed political support for aviation funding. In his next posting, as battleship captain (1918–1920), he experimented with such aircraft operations at sea as gunfire spotting and scouting. Moffett became the first director of naval aviation in 1921, working tirelessly to make aviation integral to naval operations. He was especially interested in airships, which he envisioned as flying aircraft carriers. One of them, the *Akron*, killed him in 1933, when it went down with him aboard in a storm off the New Jersey coast.

A converted collier commissioned in 1922 as USS *Langley* became the U.S. Navy's first experimental aircraft carrier. Her successful testing led the navy to convert two battle cruisers, disallowed under the Washington Naval Treaty of 1921, into aircraft carriers. Commissioned in 1927, *Lexington* and *Saratoga* were colossal ships, 888 feet long and 106 feet across, with displacements of 33,000 tons and 1,900-man crews. They also were the fastest capital ships afloat, 180,000 horsepower driving them at top speeds over 33 knots. The first American purpose-built aircraft carrier, *Ranger* commissioned in 1934, was lighter, slower, and proved an altogether less successful design, though it could accommodate a larger number of aircraft. Not until 1937 did a third heavy carrier join the fleet. *Yorktown* was an 800-foot, 20,000-ton ship with a top speed of 34 knots that could operate eighty aircraft. Within three years she was followed by two sister ships, *Enterprise* and *Hornet*, and another light carrier, *Wasp*.

The first U.S. airplane specifically designed for carrier-based operation entered service aboard *Langley* the same year she was commissioned. Built by Curtiss from a Bureau of Aeronautics design, the NAF TS-1 was a single-seat biplane fighter built of wood and fabric, and armed with a single 0.30-inch Browning machine gun synchronized to fire through the propeller. Other fighters followed, notably the series of Boeing biplanes from 1927 to 1937, as well as more specialized aircraft, like the Martin T3M torpedo bomber and scout that entered service aboard *Langley* and *Lexington* in 1927, and the Curtiss and Martin dive-bombers that began arriving in the early 1930s.

Aircraft performance improved steadily during the interwar period, with a large assist from NACA research. Improvement had less to do with radical innovation than with steady refinement of engineering practice reflected in engines of growing power, closer attention to streamlining and reduced drag, variable-pitch propellers, and wing flaps. Civilian aircraft actually led the way, but by 1937 the navy was beginning to receive its first monoplanes designed for carrier-borne operations: Douglas torpedo planes in 1937, Douglas dive-bombers in 1940, and Grumman fighters in 1940.

AMPHIBIOUS WARFARE DEFINED

Carrier-based aircraft also contributed to a remarkable reversal of prospects for success in making opposed landings on hostile shores, aided by such innovations as bombardment rockets and amphibious vehicles. In the First World War the British attempt to invade Gallipoli from the sea was a costly fiasco. Between the wars, the U.S. Marine Corps studied the problems of seaborne assault under fire, developed a suitable doctrine, and acquired the specialized vehicles required; in World War II, American forces succeeded in dozens of such operations, culminating in the extraordinary invasion of Europe in 1944. This success depended on the Allies winning complete local air control, but good tactics and machines on the surface were no less indispensable.

The Marine Corps mission had always included defense of forward naval bases. In the bureaucratic struggle for survival imposed by curtailed military budgets after the World War, the marines sought a new mission significant enough to insure an independent existence. The doctrine of amphibious warfare provided the rationale for an independent Marine Corps, just as doctrine of strategic bombing justified an independent air force. With the support of Marine Corps Commandant John A.

Lejeune in the 1920s, forward-looking young officers at the newly established Marine Corps Schools in Quantico, Virginia, began formulating the doctrine that in the mid-1930s found expression in the *Tentative Manual of Landing Operations* and verification in maneuvers off Puerto Rico.

Amphibious doctrine, like armored doctrine, outran the machines it required. Throughout the 1930s the Marine Corps sought something better than ships' boats to land troops and materiel on hostile beaches. Disappointed in its own experiments, the corps found outside help. The day of the lone inventor had not entirely passed. Andrew J. Higgins of New Orleans tried to interest the navy in his "Eureka" boat as early as 1926, though it took another twelve years to persuade the skeptics. Designed for use in bayou waters, it featured shallow draft, protected propeller, and a broad, flat bow that made it easy to beach and refloat. It needed only a retractable bow ramp to become the navy's Landing Craft Vehicle and Personnel (LCVP), the boat that brought Marines ashore under fire for the next three decades. Higgins also contributed to the development of the larger landing craft needed to bring heavy equipment ashore, designing the 50-foot Landing Craft Mechanized (LCM) that became another mainstay of amphibious landings.

A civilian inventor also conceived the third major landing craft. Living in Florida, retired engineer Donald Roebling devised an amphibious tractor for rescue work in the Everglades or after hurricanes. Propelled by its tracks both ashore and afloat, drawing less than 3 feet of water, seaworthy and close to unsinkable, the lightweight "Alligator" became the subject of a *Life* magazine story in 1940. Thus brought to military attention, its obvious potential led to its prompt adoption as Landing Vehicle Tracked (LVT). Variously modified and armed, it became a crucial factor in amphibious operations during World War II and long after.

TOWARD A NEW KIND OF WAR RESEARCH

World War II saw science harnessed to war, and institutions established to direct technological innovation toward desired ends. The problem of turning ideas into hardware, if not entirely solved, at least appeared entirely practical. The National Defense Research Committee (NDRC) and Office of Scientific Research and Development (OSRD) provided the basic framework for applied military-technological research in World War II, and universities became major participants in war-related technologically-oriented research.

Both sides of the new military-scientific partnership mattered. Military institutions able to recognize the value of research and willing to support it were only one side. By no means a minor issue, military resistance to scientific weaponeering seemed quite rational given past failures. Experience would soon show, however, that a stream of useful new devices from the laboratory would tend to overcome most qualms. The other side of the partnership, organizing researchers for the job, posed equally crucial questions. Scientists and technologists willing and even eager to do war research presented little problem. Every American war had produced them, and they appeared quickly after 1939 at merely the prospect of war with Nazi Germany. What made the eve of America's World War II different was the organization of proper means to convert offers of help into directed research leading to engineering development of actual weapons. In short, the United States in 1940 and 1941 learned how to harness research to military needs effectively.

The key was directed team research. In contrast to the chemist's war of 1914–1918, World War II has been called a physicist's war. It might better be termed a physicist's and engineer's war. Lines between scientific research and engineering development, the juncture of applied physics and science-based engineering in the service of technological innovation, never easy to draw precisely in the best of circumstances, blurred beyond definition under wartime pressure. Persistent images of lonely scientists or heroic inventors notwithstanding, team research had been gaining importance since the late nineteenth century.

Evaluating proffered inventions had been the main role of science in the Civil War. World War I produced a rough parity between an invention-rating Naval Consulting Board and a research-promoting National Research Council. In World War II the task of dealing with unsolicited inventions was relegated to a corner in the Commerce Department, while research took center stage. The result has profoundly shaped relations between American research and military institutions ever since.

The NDRC and OSRD improved or developed an extraordinary array of weapons and other products and processes useful to fighting World War II. Radar and proximity fuze topped the list of those that most decisively affected the war's course, but there were many others, ranging from operational analysis to blood plasma. What made the World War II experience unique was the speed of innovation. By 1940 the United States possessed an unmatched combination of resources in science, engineering, technology, and industry. Properly organized and directed, such resources could turn ideas into weapons quickly and massively enough to alter the course of war in progress. No other nation in World War II equaled the

United States across the entire spectrum of research applied effectively to war. NDRC and OSRD were the main reasons.

Credit for conceiving NDRC belonged chiefly to Vannevar Bush. Fresh from doctoral study in electrical engineering at Massachusetts Institute of Technology (MIT) and Harvard, he had worked for the navy on submarine detection in 1917. Back at MIT after the war, Bush proved himself an outstanding teacher, researcher, and administrator. In 1939 he became president of the Carnegie Institution of Washington and also chairman of NACA. Bush thus found himself at the heart of America's scientific establishment when war erupted in Europe, just months after news of the discovery of nuclear fission in a German laboratory. Well aware of German technical and scientific prowess, Bush and his colleagues feared the prospect of a uniquely powerful bomb in Nazi hands. Science and technology would clearly play a major role in the coming war, into which the United States must almost certainly be drawn.

Vannevar Bush (1890–1974)

The youngest of three children, Bush grew up in Chelsea, Massachusetts, a good student though often sickly. He attended nearby Tufts University, graduating in 1913 with both B.S. and M.S. degrees. Tufts made him a mathematics instructor, but Bush wanted more. He earned a joint MIT-Harvard engineering Ph.D. in 1916 and returned to Tufts as assistant professor of electrical engineering. He and Phoebe Clara Davis married the same year. They soon had two sons. During World War I, Bush invented an electromagnetic device to detect submarines, but found dealing with the navy a frustrating experience.

In 1919 Bush joined the MIT electrical engineering department as associate professor, rising to full professor four years later. He continued to invent, most notably, with his students, an electromagnetic analog computer. A strong advocate of university research in the interwar period, Bush and his associates established academic research facilities to serve both the military and industry during World War II. Bush became president of the Carnegie Institution of Washington in 1938, and chair of the National Advisory Committee for Aeronautics as well.

As World War II approached, Bush persuaded President Roosevelt to appoint him to head a reactivated National Defense Research Committee, which soon expanded to become the Office of Scientific

Vannevar Bush's Differential Analyzer. (MIT Museum)

Research and Development. From this position during World War II he promoted radar, the proximity fuze, the atom bomb, blood plasma, and a host of other innovations. His postwar report to the president, *Science—The Endless Frontier* greatly influenced future scientific research in the United States. He also did much to shape public perceptions of science's role in national defense with his best-selling *Modern Arms and Free Men: A Discussion of the Role of Science in Preserving Democracy* of 1949.

Bush set himself to mobilizing scientific and technological research. Experience in World War I persuaded him that the military technical services were too narrow to imagine genuinely new weapons; they could at best improve what they already had. The civilian side had not done much better. Screening inventions like the Naval Consulting Board was a relic of nineteenth-century thinking. Though a better idea, the National Research Council was also badly designed. As a privately funded organization, it lacked both authority and budget to impose any real direction on American science. In Bush's eyes, NACA offered a much more promising model for what in June 1940 became NDRC. Rather than building new research facilities from scratch, for instance, NDRC adopted NACA's method of research contracts for selected problems with well-established academic and industrial organizations. Bush kept administrative control of NDRC, but left technical decision-making to five research divisions: armor and ordnance, chemistry and explosives, communications and transportation, instruments and controls, and patents and inventions.

Although it would, of course, discuss service needs, NDRC would not simply accept assignments from the armed forces. Military utility and unlimited money merely defined the framework; in accord with Bush's vision, the choice of specific research problems remained in expert hands. That left the crucial decisions to science: recognizing what technical possibilities inhered in science and judging which prospects to pursue with the talent and resources available. The constant goal was weapons useful in the current crisis, the focus on applied research of immediate utility. Relying chiefly on academic research facilities, NDRC left industrial and military research groups to pursue their own ends. Industry would be gearing up for an enormous production effort, to which it would have to devote all its resources; military research would presumably continue to stress improving weapons already in use.

By mid-1941 NDRC was reaching its limits and Bush devised a larger organization that included development as well as research. The OSRD subsumed NDRC in an advisory role, joined by a new committee for research in military medicine. The five divisions of NDRC expanded into eighteen, supported by panels on applied mathematics and applied psychology that worked with any division needing help in their areas. In contrast to NDRC's support through presidential contingency funds, OSRD enjoyed a line item in the federal budget. Assured funds allowed the fruits of research to be developed into working prototypes for production. Perhaps most amazing, the entire structure was in place and operating effectively half a year before the United States became an active belligerent. The next four years would see the spectacular consequences.

7

The Climax of Mechanization: 1942–1945

◆

World War II marked the climax of another long cycle in military history. Since the early nineteenth century, armed forces had faced a flood of new and newly improved weapons. Although some reflected military initiatives, most came from independent inventors. New weapons demanded new tactics and organization, though of just what kind was seldom obvious. Industrialization and mechanization contributed to extraordinary increases in firepower ashore and afloat that outmoded traditional offensive tactics and operations, setting the stage for stalemate and the frustration of command.

In World War II further technical and tactical innovation bore fruit, yielding opportunities for generalship comparable to the Napoleonic wars. Mechanized armed forces wedded to the tactics of infiltration restored maneuver to battle and decisiveness to war. Striking differences in the course and outcome of the two World Wars reflected the success of interwar efforts to harness research to military needs. The United States was not a leader in all these developments before the war, but neither was it the laggard it had once been. And in war research it soon emerged a pacesetter.

Finding efficient means for turning research into weapons across the full spectrum of modern warfare was the crucial problem that the United States largely solved in organizing for World War II. Instead of relying on makeshift expedients, as it always had before, the nation enjoyed a well-considered and effective system for making research useful to the war effort. Nor were consequences limited to the war itself. Arrangements made to exploit research in World War II permanently transformed relations among American military, technological, and scientific institutions. Academic and industrial laboratories largely dependent on government funding have become major sources of military-technological innovation during the later twentieth century. Some aspects of this new order emerged during the nation's brief plunge into World War I, but World War II marked a coming of age.

RADAR AND THE RISE OF AUTOMATION

Electronics became the focus of support by the Office of Scientific Research and Development (OSRD), its chief research center the Radiation Laboratory at the Massachusetts Institute of Technology, commonly referred to as the Rad Lab. Electronic applications of many kinds dramatically improved techniques of integrating combat arms, controlling battle, and destroying the foe, but radar was by far the most important. All the major powers knew the principles and were working on ways to employ radar (for radio detection and ranging) in the 1920s. By 1938 the U.S. Navy had installed a prototype radar system on the battleship *New York*.

But the real contest between radar measures and countermeasures began in 1940 during the Battle of Britain. The United States soon followed Britain's lead and, thanks to OSRD and the Rad Lab, made rapid progress. At British urging and aided by the British-invented magnetron, the United States specialized in microwave radar. Over the course of the war, the MIT Radiation Laboratory produced 150 systems to serve purposes as varied as detecting enemy planes and ships, directing guns, aiding navigation, controlling operations, locating targets, and warning of attack.

Proximity fuzes were another OSRD product that mattered, well under way before the United States was formally at war. The idea was simple: Put a tiny radar set in the nose of a bomb or a shell to measure precisely when it reached a specified distance from the target; the explosive could then be reliably detonated at the set distance. When the National Defense Research Committee (NDRC) opened for business in mid-1940, the navy proposed work on radar-fuzed shells as a likely counter to enemy air attack. Merle A. Tuve, on leave from the Carnegie Institution of Washington for the duration of the war, chaired NDRC's Committee T, which oversaw the development. In August the first research contract went to Carnegie for preliminary studies. A prototype fuze for 5-inch shells was ready the month after Pearl Harbor.

Merle Antony Tuve (1901–1982)

Both of Tuve's parents were educators, his father was the president of Augustana College in Canton, South Dakota, and his mother was a music teacher there. In 1918 Tuve entered the University of Minnesota, earning a B.S. in electrical engineering in 1922. Another year's study at Minnesota secured his master's degree in physics. In 1924 he moved to Johns Hopkins University in Baltimore, where he taught physics while working on his Ph.D. Collaboration with Gregory Breit provided Tuve's dissertation topic. They established the existence of the earth's ionosphere, and determined its height and density. They also contributed to the development of modern radar, and later developed a particle accelerator.

Testing of the proximity fuze at Blossom Point Proving Ground. (National Institute of Standards and Technology [NIST])

After receiving his Ph.D. in 1926, Tuve joined the Carnegie Institution of Washington as a staff member in the Department of Terrestrial Magnetism. A year later he and Winifred Gray Whitman married. They had two children. Tuve's research centered on the evolving field of nuclear physics until 1940, when he took leave of absence from Carnegie to join the war effort. Under his direction the proximity fuze was perfected. Returning to the Carnegie Institution after the war, Tuve became interested in geophysics and radio astronomy, to which he also contributed significantly.

The Applied Physics Laboratory of Baltimore's Johns Hopkins University, which Tuve directed from 1942 to 1946, was then created to complete development. Production was under way by fall 1942, and the new fuzes, called VT (for variable time) as a security precaution, began reaching the navy before the year ended. The new shells proved to be at least three times as effective as their closest competitor, shells fused to explode at a preset time, and as much as fifty times better than shells fused to explode on contact. Proximity fuzes also enhanced the performance of such other explosive devices as torpedoes, depth charges, and rockets.

Extraordinarily successful in its own right, the VT-fuze, as sensor in a complex computer-controlled firing system, also marked a large step toward automation. World War II forged the modern military-computer linkage. Analog computers, mechanical and electromechanical, had deep roots in the practical demands of compiling such products of tedious calculation as actuarial, navigational, and (by the First World War) artillery firing tables. During World War II the still greater demands posed by fire control systems for antiaircraft defense led to further advances in automation.

Although analog computers provided most wartime firing tables, the task also prompted further development of electronic digital computers, a much speedier alternative highly successful since the war's beginning in breaking enemy codes. Computer-generated firing tables programmed into gun directors, also analog computers, dramatically enhanced the accuracy of VT-fuzed shells. In the final step, target-locating radar connected to gun director rendered human judgment, or even participation, largely superfluous.

Computers and radar stimulated thinking about system organization, first in Britain, then in the United States. Operations analysis was one result. Effectively using radar raised technical questions related to the equipment itself, in the first instance, but quickly led to other questions about organization and strategy. Systematic and quantitative approaches to problems of military tactics and strategy offered the exciting prospect of war made truly scientific. The relatively narrow focus of the war years involved NDRC and OSRD in such problems as hunting submarines and dropping bombs accurately. OSRD's Applied Mathematics Panel played an especially prominent role in these studies. Valuable as they were, the further development of operations analysis after the war would prove even more significant.

FIGHTING A MACHINE WAR

However greatly novel laboratory products may have affected the course of war, it was older weapons and tactics improved and refined that most directly decided the outcome. Better guns, tanks, aircraft, and ships made the difference. Mechanization did not eliminate masses of men and guns. Artillery still caused most casualties, and tanks without foot soldiers and guns could achieve little. Properly combined, however, and linked by field radio, balanced forces of armor, infantry, and artillery could in fact play their promised role: Capable of swift battlefield movement, they restored to ground forces potential for successful offensive action.

Technology dictated the tempo of war and its ultimate results. Equally significant, military doctrine and tactics caught up with technological change. World War II was fought with improved versions of older weapons. The tanks, aircraft, and other machines of World War II were neither new in concept nor strangers to the battlefield. Rather they were simply more capable versions of earlier machines tested in battle and improved incrementally through the interwar period, a process that continued, indeed accelerated, during the war itself. Throughout World War II the United States focused on producing large numbers of standard model tanks, notably M4 Shermans. Perhaps inferior to the best German tanks, they were also far more reliable and vastly more numerous.

Although much of prewar mechanized theory revealed itself wrong in detail, the larger vision of mechanized warfare proved valid: Nothing resembling the large-scale, long-term trench warfare of 1914–1918 recurred. Like the Thirty Years' War and the Napoleonic wars, World War II came at the end of a long military-technological cycle. The climax of the mechanical era made World War II a war of maneuver by restoring to commanders their ability to direct the course of battle to decisive conclusion. The promise of mechanical war could only be fulfilled, however, with the cooperation of air forces. Absent air support, armored forces lost much of their impact. Against an enemy superior in the air, they could scarcely function at all. In the final analysis, Allied victories became possible only because Allied air forces could achieve air superiority, all but complete. Skies swept clean of enemy aircraft meant that Allied ground forces could move without fear of attack from the air, as their foes could not.

ROCKET DEVELOPMENT

Tactical bombing in support of ground forces achieved notable success throughout the war, interdiction of enemy supply lines perhaps even more than direct attacks on enemy forces. No less successful were Allied long-range patrol aircraft equipped with radar and bombs or depth charges in defeating the German submarine threat in the Atlantic. It was the tactical and operational uses of aircraft that mattered most during World War II. The importance of air superiority became all the greater as aircraft armament improved, notably with the development of effective air-to-ground rockets.

Though the best-known rocket program in World War II was German—it produced the huge liquid-fueled, surface-to-surface V-2—NDRC and OSRD also pursued an active, if far less grandiose, program. Tactically and operationally oriented, the American program may have exerted greater effects on the course of World War II than the more spectacular German effort. Especially effective against armored forces and fortifications, unguided solid-fuel rockets became a major weapon in close air support and interdiction.

Rockets had been known for centuries, and had enjoyed periods of considerable military use. Their great advantage was lack of recoil, which held the promise of powerful explosives that could be fired from aircraft, light surface vehicles, or even human shoulders. Unfortunately, they were also notoriously inaccurate, chiefly because solid propellants had never reached an adequate level of strength, uniformity, and consistency. In 1940 these shortcomings no longer seemed insurmountable. With help from British researchers, who had begun work even earlier, NDRC had several rocket weapons under development by spring 1941.

Perhaps the best-known of the rocket weapons to attain operational status was the bazooka, a tube-launched 2.36-inch fin-stabilized rocket fitted with an armor-piercing shaped charge that allowed an individual infantryman to engage a tank. The idea had come from the fertile mind of American rocket pioneer Robert H. Goddard in World War I, but the Armistice cut development short. Goddard did not return to the project in World War II, but others completed the task. Standardized by early summer 1942, it was in the hands of troops before the end of the year. Other early projects included shipboard rocket-launched antisubmarine depth charges (Mousetrap) and barrage rockets with thousand-yard range that could be launched from landing craft.

Robert Hutchings Goddard (1882–1945)

Goddard spent his youth in Boston, returning with his family to Worcester, his birthplace, in 1898. Studious but sickly, the young Goddard did not complete high school until 1904. Four years later he graduated from Worcester Polytechnic Institute. After a year of teaching, he began graduate study at Clark University, also in Worcester, and in 1911 received his Ph.D. in physics. A research year

Robert Hutchings Goddard beside the first liquid-fueled rocket, 1926. (U.S. Air Force)

at Princeton followed before Goddard joined the Clark faculty, on which he remained for the next three decades.

By the time he became full professor in 1920, his rocket researches were well under way, inspired by his closely held dream of inter-planetary flight. Working largely alone, with help from a few trusted assistants and financial backing from several private or semipublic patrons, he designed, tested, and patented almost every feature of the modern rocket, including such fundamental features as combustion chamber and nozzle. During 1918 Goddard set aside his liquid-propelled rockets to develop and demonstrate a solid-propellant rocket launched from a hollow tube, intended to serve foot soldiers as portable artillery. Although the army project abruptly ceased with the Armistice, it served as the prototype for World War II's bazooka. Military authorities consistently rebuffed Goddard's repeated attempts to interest them in long-range liquid-propelled rockets, even after World War II began. In Goddard's eyes, the German V-2 was the missile he could have built with military support.

The focus of work in 1943 became air-launched rockets, which quickly proved effective against surfaced submarines and other unarmored vessels. Armor-piercing rockets took longer, but the 5-inch high-velocity aircraft rocket (HVAR) reached combat units by July 1944. Launched from low-flying P-47s and Typhoon fighter-bombers at 1,375 feet per second, the 6-foot, 140-pound HVARs devastated German armor trying to contain the Allies in Normandy or to counterattack advancing Allied forces.

THE AIR WAR

Teamed with other machines or systems, airplanes transformed the conduct of war on land and at sea from the heart of Eurasia to the far reaches of the Pacific. During World War II aircraft carriers replaced battleships at the center of naval warfare. Although surface engagements by no means van-ished, they became essentially sideshows. Great actions were fought be-tween fleets hundreds of miles apart; aircraft alone decided the outcome, no warship on the sea's surface ever glimpsing an enemy vessel. Especially in the Pacific war between the United States and Japan, aircraft flying from carriers largely dictated the course of events.

Five heavy carriers and *Wasp* (*Langley* had long since been retired and *Ranger* remained in the Atlantic) flying these aircraft formed the backbone of the U.S. Pacific fleet in the first year of World War II, holding the line, though just barely, until joined by the new *Essex*-class carriers. Conceived in 1939, these 27,000-ton vessels could attain speeds of 32 knots and operate ninety aircraft. The first of the class was commissioned on the last day of 1942; more important, she was followed by another twenty-three before the war ended, a rate of production that the Japanese industry could not hope to match. Nor could Japanese forces match the improved aircraft in growing numbers flying from the new carriers.

In many respects, World War II in the air came closest to war dictated by laboratory research. Major scientific-technical advances, from radar to atomic bombs, addressed one aspect or another of air war or defense against air attack. But there was more. Fortune seemed closely to favor the side flying the latest model airplane or deploying the most recent radar measure or countermeasure. Yet in contrast to the theorists of armored warfare, the prophets of strategic bombing misread the future in almost every respect. Even apart from its moral implications, air attack on enemy industry and population proved far harder to mount, more costly to maintain, and less productive of results than they had supposed.

Consistent success by fighter planes against heavy bombers emerged as a major factor in strategic bombing failures. Not until the long-range P-51 (Mustang) began to escort them could aerial attack by B-17s and B-24s be carried home to Germany. Another factor was the surprising inaccuracy of high-altitude bombing, which meant that destroying a target required bombers in much larger numbers than anyone had imagined before the war. Only after a costly war of attrition, when Axis defensive aircraft and guns had become too few to counter the still growing and now strongly escorted bomber fleets made possible by American productive capacity untouched by war, did attack from the air approach its promised power. At that point, atomic bombs introduced an entirely new factor. Explosives of such power seemed to restore plausibility to strategic bombing doctrine. Modest forces armed with nuclear weapons might well inflict the kind of damage that would end a war almost as soon as it began.

DEVELOPING THE BOMB

The military-scientific project that produced nuclear weapons by 1945 has become the paradigm of research in the service of war. It stands as a remarkable accomplishment, technological as well as scientific. The initial discovery of nuclear fission became public early in 1939. That it came from a German laboratory troubled several refugee scientists in the United States almost at once, and assumed ominous meaning with the outbreak of war later that year. Fear of a Nazi bomb long drove British and American nuclear research and development, though in fact the German project made little progress.

With a large assist from NDRC and OSRD, what became known as the Manhattan Project was fully under way by fall 1941. American research had followed the normal pattern of NDRC-sponsored work. Like so many other crucial weapon projects, it began well before the United States formally entered the war. Shortly after NDRC formed in mid-1940, it took over the small, year-old nuclear research program conducted chiefly by academic teams via contracts with selected universities.

The first question was whether or not the fissionable isotope of uranium, U-235, could be separated from uranium ore, of which 99.3 percent was nonfissionable U-238. A year's research suggested two reasonably promising large-scale methods of obtaining U-235, gaseous diffusion and electromagnetic separation. The discovery that U-238 bombarded with neutrons could be converted into a new fissionable element, plutonium, offered a third method, chemical separation. On December 2, 1942, scientific work consolidated under OSRD auspices at the University of Chicago proved that a chain reaction could be sustained. Reactor-produced plutonium could be used for a bomb.

Success had been anticipated, or at least seemed likely enough by mid-1942 to justify a major effort. In view of the project's vastly increased scope as it shifted from research and development to engineering, procurement, construction, and production, OSRD began turning the project over to the newly formed Manhattan Engineer District, U.S. Army Corps of Engineers, and the scale of effort skyrocketed. Headed in remarkable fashion by a hard-driving engineer officer, the redoubtable Gen. Leslie Groves, in three years the Manhattan Project designed, built, and operated an industrial plant that rivaled in scale the entire prewar American automobile industry.

Leslie Richard Groves, Jr. (1896–1970)

Groves grew up in army camps, the son of an army chaplain. After a year at the University of Washington and two at MIT, he graduated from the U.S. Military Academy in 1918. Further study at the army's Engineering School lasted, with only a brief interruption, until 1921. He and Grace Wilson married in 1922, and had two daughters. A decade of routine engineering assignments followed, all performed with distinction. During the 1930s, Groves graduated from the army's Command and General Staff College and the Army War College.

Smoke rising from the atomic bomb dropped in the port city of Nagasaki, Japan, August 8, 1945. (NARA)

In 1940 Groves went to the War Department as chief of the Corps of Engineers Operations Branch, directing construction of military facilities ranging from barracks to the Pentagon itself as the United States geared up for war. Denied a combat assignment when war began, Groves instead took charge of the Manhattan Engineer District and the secret atom bomb project. His managerial prowess and single-minded determination proved key elements in driving the huge science and engineering project to so rapid and successful a conclusion. He retired from the army in 1948 as a lieutenant general, then embarked on a second career as vice president of research with a division of Sperry Rand.

As part of a crash program, work on the bomb received top priority. Providing fissionable material was only part of the problem. Designing, developing, and proving a bomb under intense time pressure was something else. The task fell to the laboratory newly created for that purpose at Los Alamos, New Mexico, operated under contract by the University of California. Led by one of the university's most distinguished physicists, J. Robert Oppenheimer, theoretical and experimental physicists joined forces with chemists, engineers, and technicians to convert a recently discovered physical phenomenon into a militarily useful weapon. Officially, work at Los Alamos began April 15, 1943; twenty-eight months later a single U-235 bomb had destroyed Hiroshima, a plutonium bomb Nagasaki, and Japan surrendered.

From 1940 onward, the United States devised new forms for mobilizing science that succeeded beyond almost anyone's expectations. Relationships among science, engineering, and military institutions emerged from World War II utterly transformed. The reluctant military innovators of the later nineteenth century became enthusiastic seekers of technological novelty in the later twentieth. Perhaps more significant, they began finding money to support the scientific research and development engineering, as well as the testing and procurement, that new weapons required. Since the seventeenth century, scientists had justified their requests for state support by claiming the ability to enhance war-making potential. Generally, such claims proved will-o'-the-wisps, seeming near but rarely quite within reach. In the United States during World War II, science seemed at last to have achieved its long-professed goal.

BODIES AND MINDS

In some ways American medical achievements during World War II were the most remarkable of all. Disease, which before the twentieth century had always accounted for the vast majority of deaths in war, was all but eliminated as a significant factor in military mortality. Among the reasons for this success must be counted the work sponsored by OSRD's Committee on Medical Research, often cooperating with the National Academy of Sciences and the surgeons general of the armed forces.

Three research areas had the greatest impact: drugs, including atabrine for malaria control and the family of so-called sulfa drugs effective against many bacterial diseases; antibiotics, notably penicillin, which nicely complemented sulfa drugs in treating infectious disease; and insecticides, especially DDT, which proved enormously successful in controlling the arthropod vectors of several diseases. With few exceptions, the main emphasis was on turning substances known chiefly in laboratory contexts into mass-produced commodities. Medical innovations, like technological, mattered only insofar as they attained wide use.

Mortality from wounds as well as disease fell precipitously. In the Civil War 1 of 7 wounded soldiers died, a figure that improved to 1 of 12 in World War I. The U.S. Army in World War II suffered roughly 592,000 battle casualties, of whom 208,000 died in action or before they reached an aid station. Approximately the same ratio of killed to wounded had obtained in World War I, but the ratio of fatal to nonfatal wounds improved to something closer to 1 in 143. Much of this success derived from better techniques of frontline treatment and the rapid medical evacuation made possible by motorized transport.

But a share of the credit must also go to the Committee on Medical Research and its work on two key threats to the survival of a wounded soldier: blood loss and infection. OSRD-sponsored research helped solve both problems. It produced methods for bringing prompt and adequate supplies of plasma and other blood substitutes, as well as refrigerated whole blood, to the edge of combat. Sulfa drugs and antibiotics went a long way toward eliminating infected wounds as a cause of death.

World War II mobilized the social sciences to a degree quite unmatched previously. Continuing a World War I precedent, OSRD's Applied Psychology Panel further refined the use of intelligence and aptitude tests to determine personnel assignments. The panel also studied weapon designs in relation to user capabilities, an important step in the rise of human factors engineering. Among a host of other social science projects, two had special significance for the postwar development of their fields. Anthropologists, on the one hand, were called upon to evaluate the enemy, particularly the Japanese, and to consider how culture and personality might affect war- and peace-making. Ruth Benedict's influential postwar analysis of the Japanese psyche, *The Chrysanthemum and the Sword*, originated in her wartime research for the Office of War Information. Sociologists, on the other hand, studied American soldiers, attempting to answer questions about adjustment to army life and behavior in combat. Not only did their findings dispel some ancient military misconceptions, but also their techniques and experience strongly influenced the postwar development of sociology. The further militarization of the social sciences became a major trend in postwar America.

Ruth Fulton Benedict (1887–1948)

Ruth's father died when she was eighteen months old and she spent her early years with her maternal grandparents in upstate New York, then rejoined her mother, a librarian, in Buffalo. Graduating from Vassar in 1909, she worked first as a social worker, then as a teacher, before marrying Stanley Benedict in 1914. She discovered anthropology when she returned to college in 1919. After fieldwork in the Southwest, Benedict received her doctorate from Columbia University in 1923. She remained at Columbia for the rest of her

U.S. Postal Service stamp issued in honor of Ruth Benedict in 1995. (USPS)

career, rising from part-time teacher to become the first female full professor in political science shortly before her death.

Benedict's extremely influential *Patterns of Culture* in 1934 introduced the anthropological concept of culture as learned behavior to the American consciousness, melding it with concept of personality from psychology. Working for the Office of War Information during World War II, she sought to apply these concepts by identifying traits of Japanese mind and behavior that might be exploited to shorten the war, most notably recommending the preservation of the Emperor. Her wartime research became the basis for her classic 1946 book, *The Chrysanthemum and the Sword*, which greatly influenced postwar policy toward Japan.

8

Military Research
Institutionalized: 1945–1951

◆

In the decade and a half after World War II, the United States erected a vast if piecemeal institutional structure for turning research to military purpose. Intensifying cold war between West and East spurred this effort, but hot war in Korea was the key to recasting American research. Time brought difficult questions about assumptions underpinning that structure, and about consequences not only for military and scientific institutions themselves, but also for American polity, economy, and society. This chapter addresses the assembly of the American military-technological research enterprise after World War II.

Since 1945 military technology has become almost entirely the product of directed research. Institutionalized research began haltingly, largely because of postwar cutbacks in military funding. Research money remained tight until the early 1950s. Bolstering research and development across the board was a major goal of the post-1951 buildup triggered by the Korean War. Defense Department budgets remained at permanently higher, and rising, levels even when fighting in Korea ended. Military agencies and the Atomic Energy Commission (AEC) accounted for 70 percent of American research and development funding by the early 1950s, and well over 90 percent of federal funds allotted to campus research.

POSTWAR SCIENCE POLICY

Several aspects of postwar science policy bore the hallmarks of wartime experience. Explicitly designed as a war measure, the Office of Scientific Research and Development (OSRD) ceased operations when the war ended, though its formal termination by presidential order came only in 1947. Reputation and influence long survived closing the office. OSRD remained a model for effectively linking science and government, even in areas seemingly remote from military concerns. Much of postwar America's medical and scientific research bore the stamp of OSRD model and example.

The wartime success of OSRD's Committee on Medical Research altered the landscape of postwar American medical research. When OSRD closed, the National Institute of Health (NIH), a branch of the Public Health Service founded in 1930 to conduct research in its own laboratories, took charge of the medical committee's outstanding contracts. It soon became, in essence, the nation's central clearinghouse for medical research. Continuing its own research, NIH also administered a growing grants program directed chiefly to supporting medical schools. In 1948 Congress created the first of the special health institutes focused on a specific problem, the National Heart Institute, and NIH became the National Institutes of Health. Medical research attracted strong congressional support. The agency funded with less than $3 million in 1945 disposed of a $52 million budget in 1950, and that was only the beginning.

Postwar science policy likewise reflected OSRD influence. Vannevar Bush, OSRD's guiding genius, set the terms for the postwar debate over science policy in his widely disseminated 1945 report to the president, *Science—The Endless Frontier*. Empowered by his wartime success, Bush won universal support for his argument that the manifold benefits of science justified public financing of research through a national foundation. Congress had, in fact, long been considering how scientific research might best be organized to serve the nation's postwar needs. In short, no one doubted the value of research. Precisely what form a research agency should take became the issue, not whether it should be created.

Bush and his supporters favored something very much like OSRD writ large: federal funding for basic research with few strings attached, oriented toward supporting the best work without regard to distributing funds widely, meeting military and medical as well as more strictly scientific ends. Others primarily allied with the air force, for whom Theodore von Kármán became the leading spokesman, sought to create a full-fledged military research organization to support a permanently mobilized military establishment. Still others, who agreed with Bush on the need for a government-funded civilian research program, differed from him in preferring an organization under greater public control and less focused on the needs of elite Eastern research centers. Ultimately, they carried the day, but the struggle took five years. The dispute delayed establishment of the National Science Foundation (NSF) until 1950, and the result was a far more modest agency than either side had intended. Military research was deliberately excluded from the purview of the new foundation, as was the medical research already adopted by NIH. Through the early 1950s, NSF remained a modest funder of academic research.

Theodore von Kármán (1881–1963)

Born and raised in Budapest, von Kármán grew up in an academic household and distinguished himself as a math student at the elite Minta Gymnasium that his father had founded. After graduating from a Budapest polytechnic, he continued his engineering studies in Germany, winning a Ph.D. from Göttingen in 1908, the same year he saw his first flying machine and found a lifelong career.

A model of the X-15 rocket in the Arnold Engineering Development Center's von Kármán Gas Dynamics Facility wind tunnel. (U.S. Air Force)

Over the next two decades, first in Göttingen, then from 1913 as director of the Aachen Aerodynamics Institute, von Kármán established himself as one of the world's leading aerodynamicists. In 1928 he accepted an invitation to head the new aeronautical research lab at Caltech. Initially splitting his time between Pasadena and Aachen, he moved permanently to America in 1930, spurred by rising anti-Semitism in Germany. Moving with him were his mother and sister, Josephine, who long acted as her bachelor brother's hostess.

In the late 1930s, military funding greatly promoted research on rockets and jet propulsion. A large army contract in 1944 allowed that research to acquire its own Pasadena facility, the Jet Propulsion Laboratory. Meanwhile von Kármán and several colleagues launched the Aerojet Engineering (later Aerojet General) Corporation to build jet and rocket engines for profit. As an advisor to Army Air Force Chief of Staff Arnold, von Kármán contributed to a major 1945 report on the future of airpower as affected by rockets, missiles, and jet propulsion. He retired in 1949, but remained active as an advisor to both the air force and NATO until his death.

BASIC RESEARCH VERSUS APPLIED

The U.S. armed forces emerged from World War II with an array of research and development skills and organizations. All tended to focus on applied science, development engineering, and hardware related to the needs of their specific branch or bureau. Much ended with the war, but some survived, though not always in the same form. Military funds had also supported several laboratories with close university ties, besides those linked to the Manhattan Project, such as the Applied Physics Laboratory at Johns Hopkins University and the Jet Propulsion Laboratory at the California Institute of Technology. Many aspects of postwar science policy—a term that always includes technology policy, and often precious little science—bore the hallmarks of wartime experience.

Technologically oriented applied research had actually dominated the wartime effort. Producing better machines was not, however, the only purpose that united research and military concerns during World War II. The postwar world offered wider prospects for military uses of research, as Vannevar Bush, OSRD's wartime director, had testified. Other scientists also argued persuasively for the value of basic research.

When postwar plans for a federal agency to support basic research were delayed, the navy filled the breach with the Office of Naval Research (ONR). Shortly after OSRD began work in the summer of 1941, the navy had created its own research and development coordinating office. By 1945 that office had also taken charge of the Naval Research Laboratory and sought a statutory basis for maintaining the wartime military-science alliance. A 1946 act of Congress created the Office of Naval Research. Under its first director, Vice Adm. Harold G. Bowen, it promptly began to divert funds from canceled procurement contracts into research.

Accomplishing much with relatively modest funds, ONR proved a liberal patron of academic science in the immediate postwar period. Like OSRD, ONR adopted the research contract as its primary funding device. Accepting the inherent value of basic research, it found money for a wide range of projects without insisting they show direct links to naval needs. ONR also served as a training ground for science administrators throughout the government. Its widely admired management practices and policies helped shape NSF after 1950. Inspired by the ONR example, the other services created offices to support basic research: The Army Research Office in 1951 and the Air Force Office of Scientific Research in 1952.

The line between basic and applied research is inherently vague. The rhetorical promise of science has usually meant technological improvement or engineering development. Military research funding totaled some $10 billion between 1945 and 1965. A 1966 study concluded that over 90 percent of the work had been purely technological, aimed at the incremental improvement of existing technologies, and most of the remainder could best be described as applied or mission–oriented research. Basic science was virtually irrelevant. Although a later study contradicted this finding, the issue has never been fully resolved.

THE STRATEGIC ATOM

Its extraordinary success had promptly made the Manhattan Project a symbol of the power of research to transform society and a gauge against which to measure other organizations. It became virtually a paradigm for the military organization of applied research. Extravagant dreams of science-based super weapons seemed to have been realized, but the new age held promise as well as threat. Military-science cooperation so successful raised hopes of larger social benefits, of problems conquered, of other dreams realized. Profound consequences flowed from both deed and symbol. The future of nuclear energy headed the postwar science policy agenda and Congress carved out this area for special treatment. The reason was simple. Nuclear policy was linked too closely to national security for routine handling; in the postwar United States the bomb was the very symbol of science and technology as military resource.

Operation Crossroads, the Manhattan Project's nuclear display at Bikini in the summer of 1946, underlined the atom's military role. Yet hopes still ran high for the peaceful uses of a seemingly limitless source of power. Fear and hope alike colored the early decisions. Debate over what form the control of atomic energy should take was heated, the central issue whether control should be vested in military or civilian hands. Ultimately Congress opted for civilian control, but it was an ambiguous decision. The Atomic Energy Act of 1946 assigned the job to an independent agency, the AEC.

Five civilian commissioners, one serving as chairman, all appointed by the president comprised the commission proper. They made policy. The larger AEC organization included a headquarters staff of several divisions to oversee all aspects, military and civilian, of nuclear research and development, procurement, production, and use. Though the 1946 act mandated a civilian commission, it also imposed a strong military presence. Military Application became the largest division in AEC headquarters, its director by law a general or flag officer and its staff drawn from the uniformed services. Military influence reached much further. Ex-officers provided a disproportionately large share of AEC officials, both at headquarters and in the field. Developing and testing nuclear weapons remained one of the AEC's central functions throughout its career, a burden its successors have retained.

Following OSRD precedent, the AEC directed its funding through contracts for both basic and applied research. Operating contracts also proved most useful. Contracting for both research and operations remained the hallmark of the AEC and its successors. Formal agreements with universities enabled the AEC to support and expand a network of national laboratories, beginning with those at Argonne, Oak Ridge, and Los Alamos inherited from the wartime Manhattan Project. Industrial firms received contracts to manage engineering development operations at Sandia in Albuquerque, New Mexico, and such production facilities as Hanford in the state of Washington. The AEC added a second weapons laboratory in 1952, like Los Alamos managed by the University of California. Located in Livermore, California, it was named after its illustrious founder, the physicist Ernest O. Lawrence.

The Manhattan Engineer District itself dissolved in 1946. Most of its assets, facilities, and workers transferred to the AEC. Military members of the wartime project became in 1947 the nucleus of a new combined agency, the Armed Forces Special Weapons Project (AFSWP). Augmenting its rosters from all branches of the armed forces, it worked closely with the AEC to develop and test nuclear weapons. The Defense Atomic Support Agency succeeded AFSWP in 1957, itself to be succeeded by the Defense Nuclear Agency in 1971, and then the Defense Special Weapons Agency in 1996. In 1998 it became part of the more comprehensive Defense Threat Reduction Agency.

JETS AND HELICOPTERS

The North Korean invasion of South Korea in 1950 quickly brought United States intervention under United Nations auspices. Though largely fought with World War II weapons and tactics, including updated tanks and piston-engine aircraft, the combined effect of limited ends and restrictive terrain eventually imposed in Korea something very much like the trench warfare of World War I. In two notable respects, however, it differed sharply from the past: the widespread use of jet aircraft and helicopters. Both had long histories.

Jet aircraft first flew in combat near the end of World War II, though in numbers too small to affect the course of events. The most successful operational turbojet-powered aircraft in the air late in the war was the German ME-262, a design that in retrospect appears several years ahead of its time. Such advanced features as swept wings and automatic leading-edge slats, when brought back to the United States after the war, helped designers at North American Aviation dramatically increase performance of the F-86 (Sabre) then under development.

Entering the race to develop jet aircraft late, the United States demonstrated the airworthiness of three models but had only one in production before the war ended. That one became the Lockheed P-80 (Shooting Star), the fighter that mainly equipped air force squadrons in the Far East when war in Korea began, though two postwar designs soon joined them: the Sabre and the Republic F-84 (Thunderjet). Piston-engine ground-attack aircraft saw significant combat in Korea, but jet aircraft assumed the major burden.

F-84s armed with HVARs, the high-velocity aircraft rockets developed late in World War II, took over the main ground-support role, while F-86s fought Russian-made MIG-15s for control of the skies over the peninsula, the first aerial combat between jet aircraft. Navy piston-engine aircraft played a larger role in Korea than their air force counterparts; the navy had also begun flying jets from its carriers in the late 1940s. They included transitional aircraft such as the Ryan Fireball, a hybrid with both piston and jet engines, and the McDonnell Phantom, the new company's first design adopted for quantity production and the navy's first all-jet fighter.

Helicopters, like jets, had seen minor service in World War II, but became a major factor in Korea. Rotary-wing aircraft promised enormous benefits in their ability to rise and descend vertically, but they also presented enormous technical problems. Not until 1939, when he successfully demonstrated his model V-300, was Russian émigré Igor Sikorsky able to persuade American military observers that helicopters could be of practical use. Small numbers of the first production models entered service late in World War II; the army used them for liaison and observation, the navy for reconnaissance and air/sea rescue.

Igor Ivanovich Sikorsky (1889–1972)

Sikorsky's parents, Ivan Alexis and Zinaida Stepanovna, were physicians in Kiev (Ukraine). They encouraged their children to appreciate the arts and sciences from an early age, instilling in young Sikorsky an intellectual curiosity that marked his entire life. In 1903 he entered the Imperial Russian Naval Academy in St. Petersburg; and three years later began his engineering career in Paris. He soon returned to Kiev, joining the Polytechnic Institute. Inspired by the successes of the Wright Brothers in America and Count Zeppelin in Germany, Sikorsky studied and experimented with flight, particularly rotor lift. While working for the Russo-Baltic Railroad Car Works from 1912 to 1917, he designed more than a dozen aircraft, including the first four-engine enclosed cabin airplane, the "Grand," and the famous World War I bomber, Il'ya Muromets.

Sikorsky and his first wife divorced and she remained with their child in Russia when he fled the Bolsheviks. After the war, he immigrated to the United States, where in 1924 he and Elisabeth Semion married; they had four children. In America he founded Sikorsky Aero Engineering Corporation. Continuing his fascination with large multi-engine planes, he designed and built the long-range amphibians that helped open the way for commercial transoceanic air travel. He also retained his interest in rotor flight, producing the first

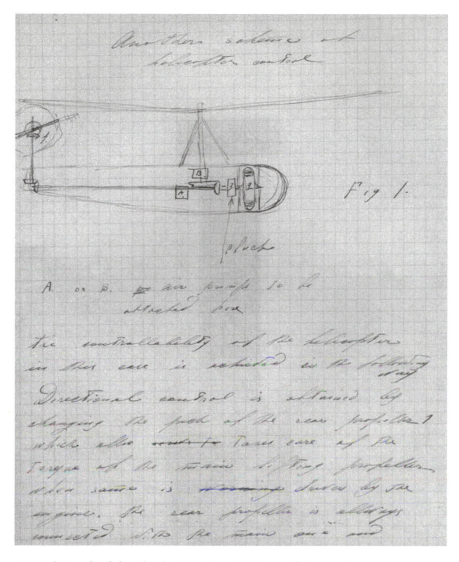

1930 design sketch by Sikorsky in his journal. (Library of Congress)

practical single-rotor helicopter in 1939. A refined version of the prewar design entered military service late in World War II.

Sikorsky, joined by Bell and Hiller, became the major helicopter suppliers for the army and navy in Korea. Primarily used in medical evacuation and rescue work, but also more experimentally in spotting for artillery, moving troops, and even raiding, they became a ubiquitous presence on Korean battlefields. They also inspired early thinking about what later came to be termed airmobility. In 1952 the army formed twelve helicopter battalions on paper, even though the required troop-carrying machines did not yet exist. Technology and tactics gestated together in the mid-1950s, leading to major changes in combat operations when the United States again went to war a decade later. In the interim, the leading edge of military-technological research lay in the direction of strategic warfare.

DELIVERING THE BOMB

The fission bombs dropped on Japan weighed 5 tons. Each of them rode to its target in a Boeing B-29 (Superfortress). The only airplane large and powerful enough for the job, the B-29 was the culmination of the long-range, four-engine strategic bomber through which prewar theorists had hoped to realize their dreams of airpower. In the war's waning months, fleets of such bombers did in fact devastate Japanese industry, as their predecessors had German. In many eyes airpower doctrine appeared to have been vindicated, though not so quickly nor so easily as its advocates had supposed. Atomic bombs confirmed to many strategic bombing's war-winning potential.

Early in 1946 the Strategic Air Command (SAC) became one of the three major combat commands into which the U.S. Army Air Forces divided itself, and it remained intact the following year when the air force attained equal status with army and navy under the newly created Department of Defense. SAC bore the imprint of its first director, Gen. Curtis LeMay, who had led the devastating strategic bombing campaign against Japan. Technologically, SAC faced a decade of transition, symbolized by the Convair B-36 with its six piston and four jet engines. Designed to carry 10,000 pounds of bombs over a range of 10,000 miles, it could accommodate a maximum payload of 84,000 pounds. During the late 1940s and early 1950s it filled the strategic bomber gap between the obsolescent piston-engine B-29 and the first all-jet heavy bomber, the Boeing B-52 (Stratofortress).

Relegated to medium bomber status by the giant B-36, the B-29 and its upgraded version, the B-50, remained by far the most numerous of SAC bombers through the early 1950s. SAC deployed its first all-jet bomber, the North American B-45 (Tornado), in the late 1940s.

Essentially a conventional bomber fitted with jet engines, it served chiefly in its reconnaissance version. Not until 1953 did SAC begin to receive a jet bomber that could fill the strategic role, though only because developing the techniques and equipment for in-flight refueling had been given high priority. Without such aid, the Boeing B-47 (Stratojet) had only a 1,500-mile radius of action, but its deployment marked the end of the line for piston-engine bombers. In 1955 B-52s began to reach operational units, and in successive modifications have remained SAC's main strategic bomber ever since.

Carrier-based naval aircraft joined the nuclear club soon after World War II ended. Two piston-engine attack bombers entered service in the late 1940s, the Douglas BT2D (Skyraider), late models of which were modified to carry tactical nuclear weapons, and the twin-engine North American AJ (Savage) designed specifically as a nuclear-strike aircraft. The navy's first jet-powered nuclear strike bomber was the Douglas A3D (Skywarrior). Design work began just after World War II for what proved to be the largest and heaviest airplane ever proposed for routine carrier use. The Skywarrior went through several versions, the first delivered in 1953 and the definitive A3D arriving in 1957. It had an operational radius over 1,000 miles with a 12,000-pound payload.

SCIENCE AND STRATEGY

Among the institutional consequences of nuclear weapons was a trend toward civilianizing military strategy. Science seemed the likeliest source of answers to the novel problems posed by nuclear war. In this context, science meant chiefly the techniques of operational analysis as practiced by OSRD's Applied Mathematics Panel during World War II. Perhaps because of what nuclear weapons implied for strategic bombing, the air force took the lead in 1946 when it contracted with the Douglas Aircraft Company of Santa Monica, California, for Project RAND (for Research ANd Development). Project RAND began modestly. Initially intended to provide advice on certain relatively technical problems, it quickly expanded to become an independent nonprofit corporation. Civilian experts assumed ever-larger roles in strategic planning. Physicists and engineers were joined by political scientists like Bernard Brodie, and by other social scientists, especially economists, who relied upon and improved the wartime techniques of operational analysis to provide quantitative answers to critical strategic problems.

Bernard Brodie (1910–1978)

Born in Chicago of Latvian-Jewish immigrant parents, Bernard Brodie grew up in tenements. He attended the University of Chicago, where in August 1935 he and fellow student Fawn McKay married. Their wedding took place on the day McKay, a Mormon from Utah, received her M.A. in English. His parents and siblings refused to attend the wedding ceremony, held in a Latter Day Saints chapel. Fawn Brodie later became an influential historian and biographer.

Poster advertising a Civil Defense training class. (Civil Defense Museum)

Brodie completed a Ph.D. in International Relations at Chicago in 1940. He taught at Dartmouth College for two years, then served in the office of the Chief of Naval Operations and, in 1945, as a technical expert for the U.S. delegation at the San Francisco conference that founded the United Nations. After the war, Brodie returned to teaching, now at Yale University, with interludes as a Carnegie fellow at the Princeton's Institute for Advanced Study and resident professor at the National War College.

In 1951 Brodie joined the newly organized RAND Corporation. He, Fawn, and their three children moved to California, permanently as it turned out. At Rand, Brodie became an influential nuclear strategy intellectual. His published analyses of the impact of the atomic bomb on international political and military affairs broke new ground and subsequent books and articles cemented his international reputation. In 1966 he returned to teaching at UCLA as a professor of political science.

Success bred imitation. The Secretary of Defense and Joint Chiefs of Staff first tried a Weapons Systems Evaluation Group, physically located in the Pentagon and commanded by a senior officer. That 1948 experiment proved a failure, unable to attract enough civilian experts. In 1956 it was succeeded by the Institute for Defense Analyses, initially a university consortium, later like Rand a nonprofit corporation. The other services meanwhile followed the air force example as well, the army with the Research Analysis Corporation, the navy with the Center for Naval Analysis.

Military budgets had dropped steeply in the five years after World War II. Relying on its monopoly of nuclear weapons and following long-standing precedents for postwar demobilization, the United States cancelled most of its wartime contracts and sharply reduced its armed forces. By 1949 the euphoria engendered by American victory in World War II had given way to an intensifying Cold War, underscored in August of that year by the Soviet success, sooner than many expected, in testing its own nuclear weapon.

The United States began to recast fundamentally its military and foreign policy, codified in NSC-68, a National Security Directive early in 1950, not long before the North Korean invasion of South Korea in June. Military budgets soared in response to the needs of troops in the field, but most of the money was not intended for prosecuting the war then in progress. Following the dictates of NSC-68, it went instead to vastly expanding American nuclear capabilities, including the crash program to develop thermonuclear, or hydrogen, bombs.

The 1950 decision to proceed with H–bomb development strongly reinforced the trend toward civilian expertise. Thermonuclear weapons exploited the explosive fission of heavy elements to ignite the fusion of light elements, in theory multiplying explosive power virtually without limit. Conventional bombs in World War II could destroy a few hundred square yards, the fission bombs dropped on Japan a few square miles. Conceivably, such limited areas might fall within the scope of military planning based on war as a rational means of settling disputes between nations. Not so hydrogen bombs. Destruction measured in hundreds of square miles, a scale more nearly akin to natural disaster than human agency, seemed to render military experience largely irrelevant. By the late 1950s, the theory of nuclear warfare and deterrence had become the almost exclusive province of civilian experts. The prospective horror of nuclear war, to say nothing of the baffling questions of fighting it, has in any event sufficed to keep it entirely theoretical.

Development, however, proceeded apace. Nuclear tests at Enewetak in 1951 and 1952 showed that fusion could be achieved. Although the air force sought and obtained a so-called "emergency capability" weapon by mid–1953, the prospective 25-ton monster, more than 5 feet across and 18 feet long, could only be carried by a specially modified B-36. Operation Castle early in 1954, again at Enewetak, proved actual bomb designs. Rapid progress in making smaller and lighter thermonuclear weapons opened up the first real opportunities for services other than the air force to join the thermonuclear club. Carrier-based aircraft would become available as delivery systems, as would land- and sea-based rocket-powered missiles.

9

Strategic Technologies Ascendant: 1952–1965

Whether or not fission bombs ended the war against Japan, nuclear weapons exerted their greatest impact on the future, an impact more relevant to institutional development than the conduct of war. Yet institutional changes matter no less than shifts in tactics, doctrine, or operational philosophies. To an unprecedented degree, the new character of American military institutions began with the Manhattan Project. The combination of nuclear weapons and long-range guided missiles has dominated strategic planning and military policy-making in the postwar world, though nuclear weapons have not been used in war since 1945. Although the kind of war for which such weapons exist has never been fought, the technology has undergone continuous development without respect to practical experience.

During the 1950s and 1960s, the strategic defense of the United States dominated military planning. Although literal defense figured in the equation, most of the money and research went to offensive weapons. Effort focused on completing the development of intercontinental ballistic missiles under air force auspices and the navy's fleet ballistic missile system. Along with already deployed long-range jet-propelled bombers, they formed what came to be called the strategic triad. Each leg of the triad was fully capable of launching overwhelming nuclear retaliation against any attacker. Together the triad gave the United States a credible deterrent, since no conceivable attack stood a chance of crippling all three legs.

THE CHALLENGE OF SPUTNIK

The United States faced a military-scientific crisis after the Soviet Union launched the first artificial satellite on October 26, 1957, and a second eight days later, this one with a dog as passenger. Two Sputniks in little more than a week shook the casual confidence many Americans placed in their country's scientific and technological prowess. One direct consequence was a new federal science agency, the National Aeronautics and Space Administration (NASA). NASA incorporated the old National Advisory Committee for Aeronautics (NACA) and a number of military rocket and space projects. Like its predecessor, it had its own laboratories but also relied on academic and industrial contractors for much of its research and development. Unlike NACA, NASA became heavily involved in engineering development. How the new space agency was to be structured provoked a struggle in some ways reminiscent of the dispute over the Atomic Energy Commission (AEC) a decade earlier, again raising questions about civilian versus military control of research. And once again, the outcome was an ostensibly civilian agency with large military participation; all NASA's early launch vehicles, for instance, were modified missiles.

Suddenly, after Sputnik, science and technology advice for the president again became a high priority. A new post, Special Assistant for Science and Technology, was established in the White House. Promptly filled by MIT president James R. Killian, Jr., it marked the first time an American president enjoyed the services of a full-time science advisor. At the same time, the Science Advisory Committee, created in 1951 in the Office of Defense Mobilization, moved to the White House as the President's Science Advisory Committee. Given a voice at the highest levels of government, science flourished over the next decade. The decision reflected still widespread beliefs about science as a source of military technology, and concerns about using it properly in the national interest. Much of the advice sought concerned nuclear weapons and missile development.

Pentagon reorganization in 1958 replaced the largely advisory Assistant Secretary of Defense for Research and Development (established in a 1953 defense reorganization) with a Director of Defense Research and Engineering who enjoyed direct authority to approve, reject, or modify all defense research projects. Reorganization also created an Advanced Research Projects Agency (ARPA, later the Defense Advanced Research Projects

Agency, DARPA) able to act promptly on special projects, especially in the areas of space and missile defense.

As a major supplier of research funds, the Pentagon exerted increasingly strong effects on the direction of research and even the structure of universities, which came to depend on such funds. This tendency became more pronounced with the passage of the National Defense Education Act (NDEA) of 1958, still another consequence of Sputnik's blow to national pride and the implied military threat. NDEA provided immense sums of money to channel students into courses of study the government deemed useful for national security, with a strong accent on science and engineering.

The success of science symbolized by the Manhattan Project could not easily be separated from the fear nuclear weapons generated. Radioactive fallout from nuclear weapons testing inspired a wave of protest during the 1950s, but qualms about science and technology, especially as linked to military imperatives, spread much more widely. President Eisenhower's 1961 farewell address articulated some of these concerns. His now-famous warning against the military-industrial complex was only part of the message. He also warned against twin dangers of science too closely entwined with government: academic research might suffer from excessive dependence upon federal support, but government policy-making might be surrendered to a scientific-technological elite. Government in this context largely meant the armed forces.

STRATEGIC MISSILES

Soviet satellites represented more than merely a blow to American pride. They also posed a clear, if not explicit, military threat. Boosters powerful enough to lift a payload to space might just as easily loft a nuclear bomb across oceans, and guidance systems able to place a satellite in orbit might well be capable of putting a warhead on target. A surprise missile attack could destroy Strategic Air Command (SAC)'s manned bombers, upon which the United States relied to carry nuclear weapons to the enemy. Without bombers the nation would be left unable to retaliate. Motivated in part by such concerns, the United States soon began to deploy its own missile force.

That such missile programs were well under way at the crucial moment, however, had little to do with SAC vulnerability. Stimulated by an exciting new technology, all three branches of the armed forces were at work on guided ballistic missiles by the early 1950s. German technologists brought to the United States after World War II by Project Paperclip gave American rocket research a major boost. Werner von Braun was only the most famous of the many German scientists and engineers who shaped the American missile and space program. Some of the earliest work in America was more scientific than military, as the reassembled V-2 rockets launched from White Sands, New Mexico, provided new and previously inaccessible data on the upper atmosphere. Although work soon moved beyond the German wartime achievements, intermediate and long-range ballistic missile programs proceeded with little urgency and many question marks.

Intercontinental ballistic missiles (ICBMs), in particular, posed formidable technical problems: Nuclear warheads, the most plausible payload, seemed too heavy, guidance systems too inaccurate, for the state of the art in the early 1950s. That changed as rockets and guidance systems improved, but the key breakthroughs came in nuclear weapons design. More efficient warheads meant lighter payloads, while the vastly greater power of thermonuclear explosions relaxed demands on guidance by the mid-1950s. These developments had several consequences.

In 1956 the army lost its longer-range missile programs, concentrating thereafter on short-range rockets for tactical nuclear weapons. Acquiring control of all long-range land-based missiles, the air force promptly accelerated its ICBM programs, Atlas and Titan, as well as Thor, the intermediate-range ballistic missile (IRBM). Activation of the first operational Atlas squadron at Vandenberg Air Force Base, California, followed in April 1958; seven months later an Atlas missile completed its full-range operational test flight, hitting a target area over 6,000 miles away. Both the single-stage Atlas and the two-stage Titan used cryogenic propellants, making them slow to launch and vulnerable to attack.

By the mid-1950s, new research had overturned the unfounded belief that solid propellants were inherently unreliable; researchers also found chemically energetic combinations of liquid fuel and oxidizer that did not require temperatures near absolute zero. Purse strings loosened by orbiting Sputniks allowed development of second-generation ICBMs—the solid-propellant Minuteman and the more powerful Titan II with storable liquid propellants—to begin without slowing Atlas or Titan. Both missiles could be protected in hardened underground silos ready for immediate launching. Successful operational test flights during 1962 brought them into the American arsenal. By 1967 SAC had added a strategic missile force of 1,000 Minutemen and 54 Titan IIs to augment its fleet of jet-propelled B-52 bombers with intercontinental range.

NUCLEAR POWER AT SEA

Nuclear propulsion was a hot topic immediately after World War II. The destruction of two Japanese cities by atomic bombs in 1945 and the widely reported demonstration of A-bombs at Bikini in 1946 brought home the power of nuclear weapons. It also spurred a broad range of speculation on other uses of nuclear power, most notably in transportation. For both the air force and the navy, nuclear power held the promise of machines with all but unlimited endurance.

Hoping to deploy nuclear-powered aircraft, the air force and the AEC sponsored two major nuclear engine programs: turbojets for manned bombers (ANP, or Aircraft Nuclear Propulsion, 1951–1961) and ramjets for unmanned bombers or cruise missiles (Project Pluto, 1957–1964). Both programs shared the key technical idea that heat from a nuclear reactor could replace heat from burning fuel. Neither program got as far as flight testing. Unworkable technology may well account for the failure of ANP, the manned nuclear turbojet program, but technical shortcomings will not so easily explain what happened to Pluto. The failure of the nuclear ramjet program seemed more a matter of competition from ICBMs and the defense bureaucracy's disdain for unmanned aircraft.

Naval nuclear propulsion, in contrast, enjoyed both technical and bureaucratic success, largely through the efforts of an initially obscure naval engineering officer, the redoubtable Hyman G. Rickover. At the outset, the key problem involved choosing a reactor design at a time when very little was known about reactors. Relying on sound engineering judgment and an intense course of instruction at the AEC's Oak Ridge Laboratory, Rickover made a long series of decisions that worked. He promised to have a nuclear-powered submarine in the water by 1955, and he did.

Hyman George Rickover (1900–1986)

When he was six years old, the young Rickover with his mother and sister left the poverty and religious persecution of Russian Poland to join his father, a tailor, in America. In 1918 he won an unlikely appointment to the U.S. Naval Academy, which had greatly expanded classes in response to World War I. After five years of sea duty as an engineer officer, Rickover returned to Annapolis in 1927, then went to Columbia University for postgraduate studies in electrical

USS *Nautilus* (SSN 571), the first nuclear-powered warship, commissioned September 1954. (U.S. Navy)

engineering. At Columbia, he met Ruth Masters, a student of international law, whom he married in 1931. They had one child. Ruth died in 1972; in 1974 he married Eleonore Ann Bednowicz, a navy nurse corps commander.

Rickover joined the submarine service after winning his M.S. from Columbia, but left when he failed to receive a boat of his own. In 1937 he abandoned the idea of command to become one of the navy's elite engineering-duty-only officers. In World War II Washington as chief of the electrical section in the Bureau of Ships, he achieved outstanding results with decisive, if not always orthodox, methods. He also learned a great deal about working with industry. The wartime experience stood him in good stead when he took charge of the navy's postwar nuclear program and, at the same time, served as head of the AEC's nuclear reactor division. A demanding leader and an astute politician, he played a key role in converting the navy's submarine force and a number of its surface ships, particularly aircraft carriers, from diesel to nuclear power. He also oversaw construction of the first civilian nuclear power reactor in the United States.

When *Nautilus* was launched in January 1954, she was the world's first nuclear-powered ship. Her power plant, a compact boiling-water reactor, could drive her either on the surface or underwater at better than 20 knots.

More important, *Nautilus* was the first true submarine; on her mid-1955 shakedown cruise she eclipsed every record for running submerged. In making the underwater passage from Connecticut to Puerto Rico in 84 hours, she averaged 16 knots over a distance of 1,300 miles. No submersible boat had ever traveled a tenth as far continuously submerged, nor had any sustained so high a speed underwater for more than an hour. Later *Nautilus* did even better, averaging over 20 knots on a 1,400-mile voyage from Key West to New London. Impressive as such figures were, they were soon surpassed, as nuclear reactors became the standard power plant for all U.S. submarines.

Nuclear reactors came to power surface vessels, though more slowly and ultimately less completely than submarines. Their advantages were simply less compelling on the surface: Endurance far greater than conventionally powered ships barely outbalanced far greater costs. Keels of the first nuclear-powered surface vessels were laid in the late 1950s—the guided-missile cruiser *Long Beach* in 1957, the aircraft carrier *Enterprise* in 1958, and the guided-missile frigate *Bainbridge* in 1959—but years elapsed before others joined them.

By the mid-1980s the U.S. Navy had ninety-five nuclear-propelled attack and thirty-seven missile submarines in commission, but only nine cruisers and four aircraft carriers. Although diesel-powered submarines by the late 1950s had achieved levels of performance far beyond World War II standards, the U.S. Navy decided to build no more; all future attack submarines would be nuclear powered. So, too, were the submarines adopted for the Polaris Fleet Ballistic Missile (FBM) system. Nuclear-propelled vessels, in contrast, remain a small portion of the surface fleet.

SEA-LAUNCHED MISSILES

The navy had pursued its own missile programs, starting with Bumblebee in 1944 and leading eventually to the Talos and Terrier systems based on short-range rockets with conventional warheads designed as surface-to-air missiles (SAMs) for air defense. System deployment began in the early 1950s, with the first SAM-armed cruisers commissioned in mid-decade. The navy also sought to deploy nuclear-armed surface-to-surface cruise missiles like Regulus. Cruise missile development began shortly after World War II with Loon, an American version of the German V-1. Like Loon, Regulus was a subsonic jet-propelled aircraft. Operational in 1954 but not deployed until 1957, Regulus had only the briefest of careers. It was overtaken by the offspring of

Germany's other major wartime missile, the V–2 rocket. By the mid–1950s it began to look technically feasible to put nuclear warheads on medium- to long-range guided missiles capable of shipboard launching. Coincidentally, the first nuclear-powered submarine had just put to sea.

In 1955 the navy formed a Special Projects Office to solve the problems of launching IRBMs from ships, and so create a sea-based nuclear deterrent. The Special Projects Office owed its existence to Arleigh Burke, the young admiral who had leapfrogged dozens of senior officers to become Chief of Naval Operations in August 1955. He strongly believed that sea-launched missiles were both technically feasible and highly desirable. Their successful development would contribute mightily to American security as a nearly invulnerable nuclear deterrent. Burke handpicked the new office's director, Rear Admiral William F. Raburn, and backed him to the hilt. When the new nuclear weapons research laboratory at Livermore, California, in the person of its co-founder Edward Teller, promised a warhead small and light enough to fit within a 900-pound payload, Raburn had what he needed. The result was the Polaris missile system

The navy designed the solid-propellant Polaris, a 15-ton missile, 28 feet long and 5 feet across, that could toss such a payload 1,500 miles. Polaris, like Minuteman, benefited from funds more freely available after Sputnik. Development began in 1956 as an extraordinarily successful crash program. The first submerged launch of a Polaris missile came in 1960. By the end of that year, the first Polaris submarine, the *George Washington*, was on patrol. The last of forty-one such boats, each with sixteen missiles, went to sea in 1967, the same year the air force completed Minuteman deployment.

Armed with missiles it could launch while submerged, a nuclear-powered submarine seemed an almost invulnerable deterrent. Together submarine and missile became the Polaris Fleet Ballistic Missile system, followed in due course by Poseidon and then Trident systems as both missiles and submarines continued to develop. Technology pursued largely for its own sake had completed what the nuclear theorists now dubbed the deterrent triad: three distinct forces—manned bombers, land-based missiles, and submarines—armed with nuclear weapons, each independently capable of inflicting unacceptable damage in retaliation against attack.

FROM SAGE TO MISSILE DEFENSE

Although most analysts judged defense against nuclear attack impossible, the explosive growth of microelectronics since World War II, particularly computing, has repeatedly seemed to put that eminently desirable goal within reach. Immediate postwar computer research and development depended chiefly on military funding. It was a pattern repeated at every critical stage in the history of computers, as exemplified in the career of the transistor, later in the development of integrated circuits. Although individual members of the armed forces, like the navy's Grace Hopper, made significant contributions, military money more often went to bridge the gap between laboratory breakthrough and commercial viability. Yet the crucial role of early military sponsorship, not only of electronics but also of a host of technologies, may vanish from sight and mind as change ramifies throughout society.

Grace Murray Hopper (1906–1992)

Grace Murray, the oldest of three children, was born and raised in New York City. She graduated with honors in mathematics and physics from Vassar College in 1928, going on to Yale for a Ph.D. in mathematics. While in graduate school, she married Vincent Foster Hopper, a professor of comparative literature at New York University. She retained his name after their 1945 divorce. Determined to aid the war effort with more than teaching recruits, she joined the navy's WAVES (Women Accepted for Volunteer Emergency Service) in 1943 as a lieutenant. The navy sent her to Harvard, where she calculated firing tables for the Bureau of Ordnance and assisted Howard Aiken's work on the first American programmable computer, the electromechanical Mark I.

Hopper remained on active duty until 1949 as a Harvard research fellow in engineering sciences and applied physics supported by navy contracts, helping complete the Mark II and Mark III electronic computers. Maintaining her navy reserve status, much of her postwar work centered on programming languages. She led the team that created the first compiler for machine-readable programming languages that used words rather than numbers and symbols. Her proudest achievement was COBOL, a ubiquitous programming language for three decades. Hopper faced mandatory retirement from the naval reserve in 1966, but Congress recalled her to active duty within a year and granted annual extensions until 1986, when she was the navy's oldest serving officer, an unprecedented forty-three years, and a rear

Colonel Howard Aiken and Lieutenant Grace Hopper with the Mark I, 1944. (NMAH Archives Center)

admiral. She lived to see her prediction that computers would shrink from room size to desktop units become fact.

Initially, costly and unpredictable computers attracted little commercial interest. Military support did not demand quick returns. Despite wartime successes and indirect benefits flowing from such scientific uses as analyzing the dynamics of nuclear explosions, the huge and fragile machines of the 1940s and early 1950s were hardly suitable for weapon systems. Yet automation, like mechanization a generation earlier, caught military imaginations. Faith in the longer-term prospects for enormously enhanced command and control mattered more than the failure-prone and hard-to-use machines immediately available.

By the 1950s such prospects appeared promising enough for large-scale application. Continental air defense provided the problem. The answer was a centralized command and control system termed SAGE (Semi-Automatic Ground Environment). Enormous mainframe computers would process data from a vast network of distant radar stations, then direct interceptors against attacking bombers. Simply building the system—including seventy-eight radar stations on the shores of the Arctic Ocean to form the DEW (Distant Early Warning) Line initiated by the Truman administration in 1952, as well as forty-six four-story air-conditioned concrete structures, each housing one of the 175-ton SAGE computers—was a fifteen-year feat of engineering.

Integrating the system proved an even greater challenge in an era of rapid technological change. With manned aircraft giving way to guided missiles or vacuum tubes to transistors, with virtually every system component changing in greater or lesser degree, often more than once, difficulties were immense and inevitable. Although declared operational in 1963 and surviving, at least in part, into the early 1980s, SAGE never achieved much reliability. Meanwhile system management was transferred to NORAD (North American Air [later Aerospace] Defense Command) at Cheyenne Mountain, Colorado. The real success of SAGE was symbolic. It paved the way for not only the antiballistic missile systems (ABMs) of the 1960s and the Strategic Defense Initiative (SDI) of the 1980s, but also for the current Worldwide Military Command and Control System. It foreshadowed their flaws as well. Inherently untestable, their ultimate capabilities must remain matters of faith. Each attempt to deploy such systems has evoked opposition from those failing to share the vision.

AIR DEFENSE AND THE ABM CONTROVERSY

A fully functioning air defense system needed more than means to detect and coordinate information. In World War II that had meant only antiaircraft guns of several sizes, but postwar development in guided missiles seemed to promise a more effective weapon against incoming aircraft. Like the navy, the army began work on surface-to-air missiles for air defense in the 1940s and began to deploy them in the 1950s, first the Nike Ajax, then the much-improved Nike Hercules. Declared operational in 1958, Nike Hercules was a two-stage solid-fuel missile that could intercept attacking jet bombers flying at altitudes up to 150,000 feet at distances of as much as 80 miles. By 1963 the army had deployed well over a hundred batteries, each with three launchers, to protect cities and military bases throughout the continental United States, and abroad as well.

Nike development continued with Nike Zeus, which metamorphosed into Nike X. Nike X in turn metamorphosed into the Spartan and Sprint antiballistic missiles in 1959. As part of the Safeguard antiballistic missile system, Spartan armed with a nuclear warhead would engage incoming warheads before they reentered the atmosphere, while the shorter-ranged but much faster Sprint missile would intercept those that got through. The system also included perimeter acquisition radars to pick up incoming missiles and missile site radars to guide the interceptors to their targets.

Primarily justified as defense of missile silos, bomber bases, and command centers against a full-fledged attack, and secondarily as defense of American cities against accidental or limited attack, the Safeguard ABM system became intensely controversial. Opposition came not only from peace advocates and others on the left, but also from such stalwart cold warriors as Jerome Wiesner, a former presidential science advisor, who feared an effective ABM system would threaten nuclear deterrence.

Jerome Bert Wiesner (1915–1994)

The son of immigrants who ran a small dry goods store, Wiesner grew up in Dearborn, Michigan, where he and his younger sister attended public school. After graduation, he studied at the University of Michigan, in 1940 earning a Ph.D. in electrical engineering. As an undergraduate, he also studied math and met Laya Wainger, a math major, who became his wife, lifelong partner, intellectual companion, and mother of their four children.

Jerome Bert Wiesner. (MIT Museum)

Wiesner's work during World War II on radar at MIT's Radiation Laboratory, the Rad Lab, and on nuclear weapons at Los Alamos, set the course of his career. He returned to MIT in 1946 as a faculty member and active participant in weapons development, culminating in his taking charge of the Research Laboratory of Electronics, the Rad Lab's successor, organized under a joint services (army, navy, and air force) contract that specified research and publication in the field of electronics, physics, and communication.

But in the late 1950s Wiesner turned from improving the U.S. nuclear arsenal to promoting policies to stem nuclear proliferation. He devoted his considerable resources to arms control and disarmament for the remainder of his life, becoming a particularly outspoken critic of the ABM as President Kennedy's science advisor. He returned to MIT in 1964, where he served successively as dean of science, provost, and, from 1971 to 1980, as president.

Negotiations between the United States and the Soviet Union rendered the issue moot. The ABM Treaty of 1972 between the two nations allowed each to deploy a single ABM system. Both did so, the Soviet Union around Moscow and the United States around a Minuteman missile site in Grand Forks, North Dakota. Although the Soviet Union continued to maintain and upgrade its system, the United States unilaterally dismantled its installed Safeguard system the day after it became operational in 1975. The missiles remained stockpiled until 1983, when they too were dismantled. The ABM Treaty provided the political reason for dismissing an antimissile system, but the introduction of MIRV, multiple independently targetable reentry vehicles, provided a technical reason by shifting advantage to the attacker, who could add warheads far more cheaply than the defender could add interceptors.

Another late product of the army's air defense development program was the Patriot surface-to-air missile, which originated in 1965 as the intended replacement for both Nike Hercules and Hawk systems that would be capable of defense against tactical missiles as well as aircraft. In carefully controlled tests, Nike Hercules also proved able to intercept missiles. The successor would build on that success. Technical difficulties and rising costs delayed the program; the army's first Patriot battalion became operational only in 1983. The heart of the Patriot system is the fire unit: a trailer-mounted phased array radar, a truck-mounted control station, and several launchers carried on semitrailers. In the 1991 Gulf War, Patriot missile batteries seemed to provide an effective defense against Iraqi Scud missiles, although later analysis raised serious questions about how effective they had really been.

10

The Impact of Vietnam: 1965–1980

◆

In contrast to basic research, which has tended to decline since the 1960s, mission-oriented and applied research have flourished. Military funds still supported the bulk of America's research and development enterprise years after the acknowledged end of the Cold War. Systematic research and development geared to meet military desires became the outstanding feature of modern military-technological development. The result has been enormous growth in the range of sophisticated gadgetry deployed on, above, and around modern battlefields.

Since the mid-twentieth century the technology of so-called conventional warfare has become increasingly unconventional. Familiar weapons and equipment have been improved, sometimes to a striking degree. Retaining their basic character, tanks and guns have nonetheless acquired capabilities far beyond anything that might have been imagined in the Second World War. Dramatically enhanced weapons accuracy, communications reliability, and command and control effectiveness have radically altered the combat environment.

Technologies barely visible in the 1940s have matured in later years to reshape battle decisively. Especially have missiles come to figure prominently in so-called conventional warfare. Equipped with ever more sophisticated guidance and control systems, they have drastically reshaped the battlefield and forced war machines to new levels of competence as well.

These weapons, in sharp contrast to strategic weaponry, have periodically been tested on the battlefield.

Perhaps the most striking changes have centered on the ever-widening use of sophisticated sensors and computers to find and attack enemy forces. In contrast to such grandiose schemes as SAGE and ABM, which sought to automate decision making, efforts to automate combat have enjoyed a measure of success. The line of descent from radar-guided, computer-directed, sensor-activated antiaircraft fire in World War II to smart bombs and precision-guided munitions (PGM) has been relatively straightforward. Working systematically and diligently, scientists and technologists have upgraded, diversified, and augmented guidance techniques, computer systems, and sensors, both individually and as system components.

McNAMARA'S PENTAGON

The systems analytic approach to defense planning pioneered at RAND moved from advice to policy-making in Robert S. McNamara's tenure as Secretary of Defense. During World War II McNamara had honed a systems approach to management and analysis while helping to plan air force operations. After the war he brought his team and the new techniques to the Ford Motor Company, successfully reviving the moribund enterprise and rising to company president just before John F. Kennedy became President of the United States.

Invited to become Secretary of Defense, McNamara introduced systems approaches and cost accounting to defense planning. He also brought civilian defense intellectuals, to many of whom such approaches were most congenial, from their consultancies and advisory panels into Pentagon policy-making positions. Cost accounting increasingly dictated, or at least decisively influenced, strategic choices, while fiscal planning and budget allocation became the chief means of structuring forces and procuring weapons. This was the special province of the new Assistant Secretary of Defense and Defense Comptroller, Charles J. Hitch, the RAND Corporation economist who had coauthored the pathbreaking *Economics of Defense in the Nuclear Age* (1960).

Charles Johnston Hitch (1910–1995)

Raised in Boonville, Missouri, where his father was associated with Kemper Military Academy, the young Hitch attended Kemper before studying economics at the University of Arizona. Graduate study in economics at Oxford under a Rhodes scholarship concluded with a 1935 master's degree. In a most unusual turn of events, Queen's College elected Hitch a Fellow. He remained at Oxford until World War II, when he returned to America and a job with the War Production Board. In Washington he and Nancy Squire married in 1942. The following year he joined the army as a private and was assigned to the Office of Strategic Services.

Hitch returned to postwar Oxford with Nancy and their daughter, but in 1948 moved to Santa Monica to chair the Research Council at the RAND Corporation. There he developed the systems analytic techniques that transformed national defense policy. In 1961 the Kennedy administration brought him back to Washington as Assistant Secretary of Defense and Comptroller, charged with overhauling the process for program and budget decisions. His reform plan, called PPBS—Planning, Programming, and Budgeting System—came to be known as "Hitch craft." In 1965 he became vice president for finance of the University of California; and three years later the university's president.

The impact of such changes was not limited to matters of policy. Traditional principles of military organization began losing their unquestioned hold. Directing armed forces increasingly came to rely on management skills rather than command authority. Young officers might find a degree in business management more useful for their careers than training in strategy and tactics. Military professionalism lost much of its once pivotal role as civilian experts in game theory, systems analysis, and other esoteric specialties reshaped strategic planning and challenged long-standing views about the functions of armed forces. Such quantitative and systematic skills, though themselves products of military needs and support, would seem to have little in common with former notions of military expertise.

Larger aspects of American society were also affected. During the 1960s the Pentagon exerted ever greater control over the direction of American economic development. What Seymour Melman termed "Pentagon capitalism" turned the defense budget into a device for central economic planning with profound consequences for the conduct and content of science, social as well as physical, and engineering. Through its placement of contracts for research, development, and production, the Pentagon altered higher education and channeled key areas of technological development along lines of military interest.

The development of the automated machine-tool industry exemplifies the way military interests may transform engineering practice, even in an ostensibly civilian setting. Beginning in the 1950s, the U.S. Air Force funded research and development in techniques for controlling machine tools automatically; that is, without direct handling by workers. Among several promising methods, the air force favored numerical control. This technique emphasized the transfer of skill from worker to machine, in contrast to other methods that left greater autonomy in workers' hands. Technological efficiency may have suffered from this choice, but it strongly promoted managerial hierarchy and control of the workforce. Such choices were nothing new. Military preferences for controlling workers, even at the expense of productivity, underwrote the expansion of scientific management, or Taylorism, at the opening of the twentieth century and promoted many of the techniques of modern management.

TECHNOLOGY IN VIETNAM

Superficially, perhaps, a veteran of World War II or Korea might have found nothing too strange on the battlefields of Vietnam: aircraft, guns, tanks, and infantry still dominated operations. Appearances, however, were deceptive. Although the basic structures of combat remained largely intact, improved and novel technologies sharply altered the conduct of warfare. Such technologies resulted directly from military-sponsored research and development, which brought to practical fruition several technologies that in World War II existed only in embryo. The most significant was the development and improvement of short-range electronically guided missiles, and the related integration of computers and electronics with mechanical systems.

These new technologies made their full-scale battlefield debut in Vietnam, and the exigencies of combat forced further development. History persuaded the United States that technological innovation backed by immense productive capacity won wars. This became the accepted formula for victory in Vietnam. Although faith in technology ultimately proved misplaced, efforts to implement it led to striking innovations in several areas. Dramatic advances along three technological paths particularly marked America's Vietnam years: helicopter operations, communications and control, and the varied technologies of detection and attack characterized as "the automated battlefield."

Helicopters became the most widely broadcast image of the American war in Vietnam, in three distinct but complementary roles: ferrying troops to and from combat, providing aerial fire and communications support, and transferring casualties from the battlefield. In 1955 the U.S. Army chose the Bell Model 204 as winner in its design competition for a utility helicopter. Bartram Kelley, who had teamed with Arthur M. Young to create the Bell Model 47, the first commercially successful helicopter in the late 1940s, headed the design team. Initially designated HU-1A (thus the nickname Huey, which stuck even when the prefix later changed to UH, and which largely supplanted its official name, Iroquois), the new turbine-powered helicopters proved faster and more versatile than the piston-engine machines they succeeded. Deliveries began in 1959 and the first Hueys reached Southeast Asia in 1962. Variously modified, they became the mainstay of army operations in Vietnam and served throughout the war. An extensively modified, heavily armed version, the AH-1 Huey Cobra, became the first purpose-designed attack helicopter to see combat.

Arthur Middleton Young (1905–1995) and Bartram Kelley (1909–1998)

Young was born to a well-to-do expatriate couple in Paris, but the family returned to a Pennsylvania farm when he was one. There he was raised and there he and Kelley became childhood friends. Their paths diverged as young men. Young earned a mathematics degree at Princeton in 1927. Determined to make his mark as a practical scientist, he surveyed the world's unsolved technical problems and decided to tackle the helicopter. He proceeded by building and testing ever-improving models.

Kelley meanwhile majored in physics at Harvard, earning first a B.A., then in 1932 a master's in physics. After teaching prep-school math for six years, he teamed up with Young on his helicopter experiments. By 1941 they were able to demonstrate a stable flying model to the head of Bell Aircraft, who bought the rights and engaged them to produce a working prototype. Between 1942 and 1945, they succeeded. The Civilian Aviation Authority granted the Bell Model 47 the first commercial helicopter license in 1946.

With the success of the Model 47, perhaps the most successful helicopter ever built, the paths of the two friends again diverged. Young left Bell in 1947 to pursue a lifelong philosophical interest in bringing the whole range of pseudosciences such as astrology and parapsychology into the realm of science. Kelley meanwhile pursued

Bell helicopter Model 30. (Smithsonian Institution)

an eminently successful engineer's career. He remained with Bell, which shifted exclusively to helicopter manufacture, and eventually became senior vice president for engineering. When he retired in 1974, two-thirds of the world's operational helicopters came from projects he had headed.

Helicopters also played an important role in revolutionary advances in communications and control. Literally above the battle but linked to it electronically, helicopter-borne commanders could observe the fighting, receive information, and transmit orders to their forces on the ground. The late-1940s invention of the transistor and its subsequent development largely financed by military money led to communications gear dramatically reduced in size and weight over what the soldiers of World War II and Korea had known. A network of FM and VHF radio and microwave systems extended from squad to headquarters. Microwave and tropospheric systems linked stations throughout Southeast Asia, and the satellite system inaugurated in 1966 connected commanders in Vietnam to their civilian superiors in Washington. Echoing the navy's early-twentieth-century experience in putting radio on its ships, the expanded communications capabilities of the 1960s proved a mixed blessing: A more secure chain of command tended to restrict the initiative of subordinate commanders.

The combination of helicopters and an effective communications network that assigned specific frequencies to medical use continued the twentieth-century transformation of combat and emergency medicine. As many as 140 helicopters were assigned exclusively to the air ambulance role and one might be on the scene within minutes of a request for help. Although the percentage of deaths among those admitted to a medical facility remained nearly the same in Vietnam as in Korea (2.5 versus 2.6), the percentage of deaths per combat injury improved significantly, from 26.3 in Korea to 19 in Vietnam. And despite the tropical environment, disease never became more than a minor problem. Speed of transport meant fully equipped and even specialized hospitals were seldom as much as an hour away. The old hierarchy of medical evacuation, from battalion aid stations through divisional clearing station to rear-echelon hospital now seemed obsolete.

TOWARD AUTOMATED WAR

Military support for microelectronics and computers generated other technological capabilities that had major impacts on war fighting. The same extraordinary expansion of electronic computation that stimulated such strategic applications as SAGE also opened new opportunities for using computers in battle. In 1959 the Institute for Defense Analyses, a civilian-staffed Pentagon think tank, had assembled a panel of top scientists, the Jason Summer Study Group; it survived as IDA's Jason Division to provide periodic assessments of key problems in military technology. For summer 1966 the problem came direct from Secretary of Defense Robert S. McNamara, who hoped to find an alternative to the costly and seemingly fruitless bombing of North Vietnam.

McNamara proposed a "fence" against North Vietnamese infiltration, partly a physical barrier of barbed wire, watchtowers, and mines, but also an electronic barrier making extensive use of the new technology to locate, track, and target the enemy. A favorable response from the Jason study confirmed McNamara in his action. Despite military opposition—critics not only deemed the plan unworkable, they feared funding such ideas might also undercut more urgent needs—McNamara went ahead. To do so, he formed the deliberately misnamed Defense Communications Planning Group (DCPG). Army Lt. Gen. Alfred D. Starbird, the hard-charging former chief of the Atomic Energy Commission's military affairs division, among other science and engineering assignments, headed the new group. During its five-year existence, DCPG did much to refocus major portions of the nation's military research and development effort, in the process sparking a far-reaching transformation of conventional warfare.

If the work of DCPG could not avert American defeat in Vietnam, it may nonetheless rank, according to Paul Dickson, with the Manhattan Project as an instance of scientific-technological ideas rapidly converted to revolutionary new or radically improved military technology, if not of decisive result. Gen. William C. Westmoreland, the commander of American forces in Vietnam, could by 1969 envision future fields where "enemy forces will be located, tracked and targeted almost instantaneously through the use of data links, computer assisted intelligence evaluation and automated fire control."

Although the flood of DCPG innovation had yet to crest when Westmoreland spoke, U.S. forces in Vietnam had already begun to deploy key elements of the automated battlefield. McNamara's original plan as modified by Jason called for two distinct but complementary efforts: The first would place remote electronic sensors to direct air strikes against traffic along the Ho Chi Minh Trail in Laos, and thus restrict or prevent the movement of North Vietnamese troops and supplies southward; the second required building a more conventional fortified line, a barbed wire fence with guard towers, mine fields, and the like, supplemented by electronic sensors, to block infiltration through the demilitarized zone. Crucial developments centered on sensors and munitions.

A major part of DCPG work involved sensors, devices able to detect certain kinds of physical data, convert them to electronic signals, and transmit the results to waiting receivers. All remote sensors shared an electronic logic circuit, radio transmitter, and battery, but differed in their detection units. The most common were seismic and acoustic: The first type sensed ground vibration from footfalls or passing vehicles, the second sound from the same sources. Other remote devices detected magnetic anomalies, interruptions in self-generated electromagnetic fields, chemicals from human bodies or truck exhausts, heat through infrared sensors, or movement via small ground radar sets.

IGLOO WHITE

The various devices reached the battlefield by air, dropped from low-flying aircraft, such as the navy's two-engine, propeller-driven Neptune specially modified for the purpose. Other aircraft like the EC-121, a modified four-engine, propeller-driven Lockheed Constellation packed with electronic gear, monitored transmission from the sensors and relayed the data to a special facility established in Thailand, the Infiltration Surveillance Center. The entire system of remote sensors, relay aircraft, and surveillance center received the code name Igloo White and the bland cover name of Communications Data Management System.

Detection led to attack from the air. The surveillance center directed aircraft to the appropriate zone and guided their attack runs on the basis of computer-analyzed data, continuously updated. Even bomb release was often computer-controlled. Experience showed that jet fighter-bombers like the McDonnell F-4 (Phantom II) were too fast for the job. Older jet bombers like the Martin B-57 (Canberra) did better, as did helicopter gunships, but the most successful truck-killers proved to be converted turboprop transport aircraft, the four-engine Lockheed C-130 (Hercules). Introduced to combat in the late 1960s as the AC-130, code named Pave Spectre, it raised battlefield automation to a new level of ferocity.

The AC-130 carried its own elaborate sensor systems. The fourteen-man crew monitored data from a radar able to pick up truck ignitions, laser range finder and target designator, low-light level television camera, imaging infrared device, and ground target radar; it was also equipped with electronic countermeasures gear and a digital computer fire control system. Its armament was no less powerful: Vulcan 20-mm multibarrel Gatling guns and 7.62-mm miniguns, Bofors 40-mm automatic guns, and, in later models, a 105-mm howitzer. Shortly after its introduction, according to the air force, one AC-130 destroyed sixty-eight trucks in a single hour. But technological prowess did not alter the war's outcome. Even the destruction of over four-fifths of the supplies moving south could not prevent Vietcong guerrillas and main force units of the People's Army of Vietnam from continuing to wage war.

Failure of interdiction reflected misdirected policy, not technical shortcomings. Remote sensors themselves proved so successful that they were widely adopted for a variety of tactical uses, largely forestalling the proposed McNamara fence. Spread around fire bases, for instance, they helped spot approaching enemy forces and provide enough information for defenders to call in artillery fire on map coordinates, fire support from helicopter gunships, or bombing runs by fixed-wing aircraft. These techniques scored a notable success in the 1968 Marine defense of their isolated base at Khe San against a much larger North Vietnamese Army force.

Novel or much enhanced detection devices were not limited to remote sensors. Ultimately, the most widely used was the starlight scope, deployed in sizes ranging from rifle-mounted instruments to aircraft-borne long-range devices. Magnifying starlight or moonlight as much as 50,000 times, even the smallest version could allow a soldier to find a night target 400 yards away. Low-light level television cameras and thermal-imaging night vision devices joined new types of radar in helping target enemy forces.

SMART BOMBS

Conventional bombs, "iron" or "dumb" bombs in the new lexicon of military technology, remained the mainstay of aerial attack in Southeast Asia, improved aerodynamically and explosively but fundamentally unchanged from those used a generation earlier. One relatively minor device of the World War II era, however, the cluster bomb, in Vietnam achieved far greater currency in a wide range of new forms that took advantage of great advances in the technology of both materials and controls. The key idea was spreading a bomb's effect by packaging its payload in submunitions, which could be more or less widely dispersed by explosive or aerodynamic forces. Such payloads included several kinds of antipersonnel or antiarmor bomblets, mines, and, toward the end of the war, fuel-air explosives. Other bombs featured more radical changes.

Munitions research and development formed the second major area of the Defense Communications Planning Group's efforts to create an automated battlefield. Innovations in bombs and missiles, like those in sensors, included the improvement of old technology as well as the development of new. Outstanding among the several products of this enterprise were so-called "smart bombs." What made them smart were sensors linked electronically through microchips to aerodynamic control surfaces that could adjust the falling bomb's flight path toward its designated target. Two types of smart bombs made their debut in Vietnam, guided either by laser or by electro-optics.

The first generation of laser-guided bombs (LGBs), those used in Vietnam, were simply standard bombs to which guidance and control units were bolted. A sensor in the nose detected a spot of laser light projected on the target either by the attacking aircraft itself or by some other agent such as a soldier on the ground or a spotter plane. Although more complex and costly than laser-guided bombs, electro-optical guided bombs like the Walleye had the considerable advantage of taking care of themselves once released from the aircraft. The weapons officer on the attacking plane locked a television image of the target into the bomb's computer. When the bomb dropped, the computer compared the stored image with images received from the bomb's nose television camera as it fell, issuing signals to move the bomb's flight controls so the images matched. Accuracy multiplied.

HOMING MISSILES

Vietnam became the scene of the first large-scale display of other smart weapons as well, those missiles that came to be termed precision-guided munitions. Though efforts to devise guided weapons began as early as World War I, practical designs emerged only in the mid-1950s. Unlike such World War II rockets as the bazooka or HVAR, which were on their own after firing and followed purely ballistic trajectories, the new missiles could receive commands or generate their own data and alter their courses in flight. Precision derived from the same combination that made bombs smart, sensor-guided computerized control systems. Missile guidance took three major forms: passive homing, command, and active homing.

The earliest successful guided missiles operated against easy-to-spot targets like other aircraft or ships at sea, where relatively simple infrared or radar passive homing systems provided adequate guidance. Engaging more difficult targets required further development. By the mid-1960s, when reliable small rocket motors and solid-state circuitry made semiautomatic guidance possible, ground targets could be successfully engaged. External systems of one kind or another using eyesight, lasers, or radar designated the target and fed through microchips to provide the commands that guided the missile to its target. Toward the end of America's war in Southeast Asia, missiles carrying their own sensors, computers, and guidance logic could be launched and left to find their own targets.

Passive homing missiles began joining the American arsenal in the early 1950s. First was the air-to-air Sidewinder with infrared guidance that allowed it to home on the heat from jet engines; steadily improved in detail and expanded in capabilities over the next three decades, Sidewinders became the most widely deployed of all air-to-air missiles, largely because its inherently simple design allowed endless modification. The original design was the handiwork of William B. McLean, a practical physicist assigned by the navy's Bureau of Ordnance to the Naval Ordnance Test Station (NOTS) at China Lake, California, where he became technical director. The navy intended NOTS, established in 1943, to become a source of innovation. Applying the lessons of wartime research, the unique organization united military command with civilian technical direction.

William Burdette McLean (1914–1976)

Born in Portland, Oregon, the son of a Protestant clergyman who stressed self-reliance and a mother who taught him needlework before kindergarten, young McLean grew up with a knack for building what he needed. He graduated from Cal Tech in 1935 with a B.S. in physics. He later earned an M.S. in 1937 and a Ph.D. in 1939. A two-year postdoc at the University of Iowa followed. World War II brought him to the Bureau of Standards, where he designed arming devices for proximity fuzes. In 1945 the Bureau sent him to NOTS for a two-month visit that became a twenty-three-year career.

Work on aiming systems for air-to-air rockets convinced McLean that they would have to be self-guided. He preferred a heat-homing system for its inherent simplicity and persuaded NOTS management to let him go ahead, though the project had no high-level authorization. That came in 1951, when William S. Parsons, one-time Manhattan Project weaponeer and current Bureau of Ordnance deputy chief, saw a demonstration and ordered support. The team successfully fired a prototype in 1953. Called Sidewinder after the rattlesnake that also located its prey by heat, the missile joined the navy in 1956. McLean had already left the project to become NOTS Technical Director. After Sidewinder, he worked chiefly on space reconnaissance and weapons technology. McLean left NOTS in 1967 to become director of the Naval Undersea Center in San Diego.

Sidewinder became operational in 1955, the same year as the larger, longer-ranged Sparrow, its air force competitor. Relying on radar to find its target, Sparrow was the product of a more conventional development process. Although it was more complex, less reliable, and far more costly than Sidewinder, Sparrow nonetheless attained considerable longevity. Eventually the air force shifted to Sidewinder, but an outgrowth of Sparrow technology became the first successful antiradar missile. The air-to-ground Shrike homed on enemy air defense radar and thus helped reduce the threat to aircraft from surface-to-air missiles.

Air-to-air missiles and antiradiation missiles presented relatively easy guidance problems because their targets stood out clearly in largely uncluttered surroundings. Relatively easy targeting also tended to be true for missiles used at sea against ships. Not so for most ground targets, which had to be located and tracked amid often confusing backgrounds. Outstanding among first-generation command-guided missiles was the TOW (tube-launched, optically tracked, wire-guided) antitank missile. After launching this short-range missile, a soldier had to keep the target in the crosshairs, any movement automatically feeding control commands through the wire to the missile. Certainly not the perfect weapon, TOW nonetheless far outclassed any other infantry antitank weapon and has been deployed by many armies in vast numbers since its introduction in the mid-1960s.

Active homing missiles represented the second generation of guided missiles. The first successful fire-and-forget missile was the Maverick. Earliest versions relied on a television camera in the missile's nose to maintain contact with the target; accordingly, it was strictly a daylight weapon. Later versions relied on laser guidance and, most recently, imaging infrared sensors. Whatever the guidance system, the attack began with the pilot or weapons officer lining the target up in his crosshairs, locking it in, then releasing the missile to proceed on its own. Mavericks easily achieved direct hits in 88 percent of their test firings, and their combat performance was not noticeably inferior.

11

A New Era in Warfare?: 1980–2000

◆

In the rarefied atmosphere of strategic warfare, the theoretical world focused on the Soviet Union and preoccupied with nuclear weapons and long-range missiles, old habits died hard. American military policy vis-à-vis the Soviet Union continued to rely chiefly on retaining technological superiority through robust government-supported research and development programs. Despite decades of criticism centered on the extravagant costs and too-frequent unreliability of ever more technically sophisticated weapons and weapon systems, permanent technological revolution worked, or at least appeared to. The very spiraling costs that distressed the critics received credit by many analysts for helping bankrupt the less-efficient Soviet economy and toppling the Soviet polity.

In the more practical world of actual war fighting, the Vietnam experience persuaded many of those responsible for American military planning that superior technology—narrowly defined as machines with higher speed, weapons with greater range, electronics with greater power—was not in itself a viable policy. Uncontested technical superiority in Vietnam had clearly failed to win victory. Even before Vietnam, some critics had proposed linking technological development more closely to combat requirements, less to abstract ideas of better technology. In the postwar years, the armed forces individually and in combination began rethinking their attitudes toward technological change for the battlefield.

New attitudes did not slow the pace of technological change, but did change the ways that at least some, usually younger, officers responded to novelty and sought to revise American approaches to the conduct of war. During the last quarter of the twentieth century, the armed forces faced new challenges from the continuing computer revolution, now driven by market forces rather than military sponsorship. The further expansion of orbital satellite capabilities in communications, navigation, and other roles also included a growing civilian component. Perhaps the most exclusively military developments centered on precision-guided munitions and stealth technology. The range of advanced technologies came together to spectacular effect in the 1990s, when conflicts in the Middle East and the Balkans demonstrated that the end of the Cold War did not mean the end of war, though it might well mean another military revolution in the making.

HINDSIGHT AND TRACES

Escalation of war in Vietnam put pressure on military budgets and led to questions about the payoff on military funding for basic research, some $10 billion between 1945 and 1965. One result was Project Hindsight, a retrospective study of weapons systems deployed as of the mid-1960s to decide just how much basic research had contributed to their success. Thirteen teams of scientists and engineers picked twenty systems then in use, tracked their development histories over the preceding twenty years, and concluded that science had played a minor role. Over 90 percent of the work had been purely technological, and most of the remainder could best be described as applied or mission-oriented research. Basic science was virtually irrelevant.

When reported in 1966, such findings threatened the widely accepted rationale for military research funding. At least since the end of World War II, Vannevar Bush's claim that basic research in science underlay technological progress had gone largely unchallenged. That claim, in turn, underlay the scientists' justification for public support. If basic research played no significant part in advancing military technology, as Project Hindsight suggested, then science lost its claim to military resources.

Critics of these findings objected that twenty years could be too short a time for basic research to make itself felt in technological advances. Practical results might well take longer to appear. Equally problematic seemed other questions. Were the systems chosen for study the kind likely to display the results of basic research? Could a study focused on

incremental improvement—almost a definition of engineering development—adequately assess the impact of major breakthroughs?

Such questions, combined with deeper concerns about public support
for science, led the National Science Foundation to sponsor a counter
study called Project TRACES (Technology in Retrospect and Critical
Events in Science). Researchers at the Illinois Institute of Technology
conducted the study under the direction of Francis Narin, a leader in the
field of technology evaluation. Concentrating on major innovations and
extending the historical analysis to fifty years, the 1968 report found that
academic research did, in fact, provide the foundations of technological
breakthroughs. Curiously, neither Hindsight nor TRACES saw fit to make
use of historians despite the obvious historical questions they asked. Both
studies relied on the quantitative analysis that had come to dominate the
social sciences and defense policymaking and left little room for the traditional, more qualitative, skills of historical research.

Although the issue has never been fully resolved, government support
of basic research declined. The armed forces became noticeably less willing
to support basic research by the end of the 1960s, an inclination strongly
reinforced by Congressional action. Controversy about science tarnished
by military funds sharply increased in step with protests against the Vietnam
War. In 1969 the so-called Mansfield Amendment prohibited military funding for research not directly related to military needs. Consequences were
relatively minor. A Pentagon review of outstanding contracts found that
only 4 percent could not meet the standard of military relevance.

MISSILE DEFENSE REDUX

Strategic weapons were no less dynamic than conventional in the last
quarter of the twentieth century. Each leg of the deterrent triad received
dramatic upgrades, as symbolized by the MX land-based missile in the
1970s, the Trident fleet ballistic missile system in the 1980s, and the B-2
bomber with its extensive incorporation of stealth technology in the 1990s.
Both land-based and sea-based missiles acquired multiple warheads, which
could be targeted independently with ever-increasing accuracy. Each new
generation of weapon systems improved on the performance of its predecessor, but each also demanded huge increases in funding and often posed
intractable deployment problems.

Technological enthusiasm may have reached its pinnacle in a proposed
shield against nuclear attack, the so-called Strategic Defense Initiative (SDI)
of the 1980s. Initially dubbed Star Wars in 1983 when President Reagan

announced his plan for a shield over America, it became SDI a year later with the formation of the Strategic Defense Initiative Organization. Star Wars revived the dream of missile defense, moribund since the mid-1970s, when Congress ordered the air force to close the first and only Safeguard site immediately after it became operational. By the early 1980s, the perceived threat of a growing Soviet nuclear arsenal and the promise of new technology revived the program.

One of the most exciting prospects, vigorously promoted by Edward Teller and his associates at the Lawrence Livermore National Laboratory, was the so-called x-ray laser. A space-based weapon powered by x-rays generated in a nuclear explosion that would project an intense laser beam onto enemy missiles as they were launched, it offered an exciting alternative to antimissile missiles. Livermore became a major SDI contractor. Space-based directed-energy weapons like the x-ray laser would join kinetic energy weapons (interceptor missiles and electromagnetic launchers [rail guns]) to provide the teeth of the system. Other areas of research addressed the functions of surveillance, tracking, and evaluation of targets (ground- and space-based sensors); and of integrating the various elements of the system (computers, software, communications).

Edward Teller (1908–2003)

Growing up in a middle-class Jewish family in Budapest, Teller completed his secondary education with top grades in physics and math. From 1926 to 1933 he studied in Germany, with a first degree in electrical engineering from Karlsruhe and a Ph.D. in physics from Leipzig (1930). Hitler's rise in 1933 cut short Teller's postdoctoral work at Göttingen. He spent a year as a Rockefeller fellow with Niels Bohr in Copenhagen, where he and Mici (Maria Augusta) Harkanyi married in early 1934; they had three children. After a year's teaching in London, Teller became professor of physics at Washington's Georgetown University. The Tellers became naturalized citizens in 1941.

Meanwhile, alarmed by the German discovery of nuclear fission in 1939, Teller joined the American atomic bomb program, ending at the secret Los Alamos Laboratory. After the war, he resumed his academic career at the University of Chicago, but retained his close connection with nuclear weapons research. Between 1949, when the Soviet Union tested its first A-bomb, and 1954, when the AEC stripped J. Robert Oppenheimer of his security clearance, Teller the scientist was increasingly overshadowed by Teller the nuclear

Edward Teller in 1958 as Director of Lawrence Livermore National Laboratory.
(Lawrence Livermore National Laboratory Library)

weapons advocate as he campaigned for an accelerated H-bomb program and lobbied for a second nuclear weapons laboratory. In later years he vigorously promoted the Plowshare program for peaceful nuclear explosions and several technologies for antimissile defense.

Controversial from the outset, SDI threatened arms control in general and the 1972 ABM Treaty in particular, as well as the system of mutual deterrence that had helped prevent nuclear war since the 1950s—if it worked. That, as many critics pointed out, was a very large question. Quite apart from the host of technical problems on the very forefront of current capabilities, for many of which no clear-cut approaches yet existed, the system as a whole would be fundamentally untestable. Individual components might be shown to work, but only a successful defense against massive attack by nuclear-armed missiles could prove the fully integrated system of hardware and software. It would be, as more than one critic observed, a shield of faith.

After ten years and billions of dollars the original idea of defense based on x-ray lasers vanished. Perhaps the last battle of the Cold War, SDI was scaled back to a modest research program in the 1990s. Proponents claim that SDI helped end the Cold War by posing a threat too costly for the weaker Soviet economy to meet. But that after-the-fact justification is belied by the ongoing dream of missile defense, which refused to die even as its latest reincarnation in the form of a new antimissile missile continues to display the same shortcomings as its predecessors. With the end of the Cold War and the cessation of full-scale testing, nuclear missiles themselves threaten to become faith-based systems. America's massive stockpile stewardship program instituted in the 1990s relies on computer simulations, systematic inspection of stockpiled weapons, and regular replacement of components. But the skills of the weapon designers were intimately linked to hands-on testing, and an aging corps of experienced practitioners no longer has field laboratories in which their replacements would be trained.

SATELLITES OF WAR

A far more successful military use of space followed from the deployment of a large and growing force of orbiting satellites to collect and transmit globally a wide range of data. Military satellites have come to perform functions as varied as early warning against missile attack, surveillance, reconnaissance, communications, navigation, meteorology, and geodesy.

They proved to be major factors in maintaining nuclear stability during the Cold War, though they also raised troubling questions. What would happen to military forces that depended on information from orbit if electronic links failed, whether caused by equipment malfunction, hostile action, or some other mishap? Of even greater concern was the disruption of command and control of nuclear forces, with all that portended for the viability of deterrence.

Soon after Sputnik, the Eisenhower administration in 1960 initiated an early warning space system called MIDAS (Missile Detection and Alarm System) that relied on satellite-borne infrared sensors in high polar orbit to detect the hot exhaust gases from launched missiles. Only modestly successful, it was succeeded in the early 1970s by the much more capable Defense Support Program (DSP), which put its satellites into geosynchronous orbits. Declared operational in 1973, the system performed even better than expected. During the 1991 Gulf War, DSP provided crucial advance warning of Scud ballistic missile launches against coalition forces and Israeli targets.

Another system originating with the Eisenhower administration, code-named Vela Hotel, was designed to detect nuclear detonations from space. Intended to monitor the Nuclear Test Ban Treaty, the first pair were launched in October 1963, just after the treaty went into effect, followed by four more in the next two years. They proved to be so successful that the last four of the ten planned were never launched. Equally successful, reliable, and long-lived were the six advanced Vela satellites launched later in the decade. The last of them was deliberately turned off in 1984, having exceeded its six-month design lifetime by almost fourteen years. All in all, Vela may have been the air force space program's greatest success.

The Eisenhower administration also initiated surveillance satellite programs, one of the most successful being the first, the highly classified Project Corona, the public face of which was NASA's series of Discoverer satellites. Designed to return exposed film to earth in reentry capsules snagged by high-flying aircraft, the first attempt at a Corona satellite launch was early in 1959, but the first successes came only with the thirteenth and fourteenth launches in August. More than a hundred missions followed until the program ended in 1972.

Alongside film-based reconnaissance from orbit developed satellites that relied on digital cameras to transfer their data to earth electronically rather than physically. Initially, film images tended to be of higher resolution while digital images covered wider areas. Electronic-transmission photoreconnaissance satellites began with the 1960 launch of the air force's SAMOS I (Satellite and Missile Observation System). Mixed results led to the program's relatively quick demise, but later programs were developed, notably the Key Hole series of photoreconnaissance satellites, the first of which, KH-11, was launched at the end of 1976. Radar imaging from orbit joined optical imaging with the 1988 launch of Lacrosse I, which transmitted digital data to ground stations. From the early 1970s onward, seeing from space was complemented by hearing from space, with orbiting antennas picking up electronic emissions of all kinds.

Among the most significant innovations in military technology since World War II were communications satellites, both military and civilian. From small beginnings in the 1960s, they have come to dominate long-distance global message traffic. After several experimental satellite programs, the Initial Defense Communications Satellite Program became operational in 1968, bringing Washington into nearly instantaneous communication with Vietnam. Subsequent programs, notably the Defense Satellite Communications System from 1982 onward and the Milstar Satellite Communications System from 1994 onward, have expanded the reach and reliability of military communications. Reliable and nearly instantaneous global telecommunications have transformed the military chain of command, a mixed blessing for commanders in the field. Just as radio vastly increased the coordination of forces in World War II, the newer forms of telecommunications brought field commands still more firmly under central direction.

Both the United States and the Soviet Union deployed satellite systems for a variety of other useful military functions. Several of these systems, or their civilian counterparts, can provide valuable information for nonmilitary purposes as well, though usually at lower resolutions. Navigational satellites have been scarcely less important than communications satellites in their effects on military operations. The navy launched the first navigational satellite, Transit, in 1960, primarily to provide updates for the inertial navigation system of Polaris submarines. The 1973 launch of a NAVSTAR (Navigation Satellite Timing and Ranging) satellite initiated what has come to be known as Global Positioning System (GPS), which became fully operational in 1995. Accuracy measured in millimeters allows commanders to see the exact location of their units at any moment, removing much of the fog of war. Civilian applications are equally momentous.

The obvious importance of weather and accurate ranging to military operations underscores the significance of meteorological and geodetic satellite systems; like communications systems, they have found burgeoning civil as well as military uses. Both weather and mapping have become joint military-civilian programs. The independent Defense Meteorological Satellite Program initiated in 1962 merged with Department of Commerce and

NASA programs in 1994. No independent military geodesy and mapping satellite program ever existed, but data from photoreconnaissance satellites and NASA's Landsat program filled the gap, and the Defense Department has since 1998 gone into partnership with NASA on the next phase of Landsat development. All of America's military space systems—early warning, reconnaissance, communications, navigation, and weather—were integrated into the planning and operations of Operation Desert Storm in 1991.

DESIGNING FOR COMBAT

Perhaps one of the most disturbing statistics to emerge from the war in Southeast Asia involved air combat between jet fighters. In the Korean War, American F-86 Sabres had achieved an air-to-air kill ratio of 10 to 1 against MIG-15s. What made this success so striking was that on paper the MIG was better than the Sabre in almost every technological criterion. A young air force officer, John R. Boyd, pondered this paradox after the war. He concluded that the F-86, though out-climbed, out-turned, and out-accelerated by the MIG-15, had two decisive advantages: A bubble canopy allowed its pilot a much better awareness of his situation and its hydraulically boosted flight controls enabled him to shift between maneuvers more quickly.

John Richard Boyd (1927–1997)

Boyd grew up fatherless in Erie, Pennsylvania. He enlisted in the army in 1945, serving in the occupation of Japan and then going to college on the G.I. Bill. Enrollment in ROTC led to flight training after graduation. Outstanding as a fighter pilot in Korea, Boyd won assignment to the air force's dogfighter school in Nevada, where his reputation grew. His 1960 manual, "Aerial Attack Study," reshaped aerial combat doctrine around the world. After studying engineering at Georgia Tech in the early 1960s, Boyd further refined his ideas in a mathematical theory of energy maneuverability that provided a remarkably effective way to derive engineering specifications from preferred tactics.

Boyd's application of theory during a stint at the Pentagon in the late 1960s and early 1970s greatly influenced the shape of two major air force fighters, the F-15 and the F-16. In retirement Boyd read widely in a variety of fields and formulated his theory of the crucial links among observation, orientation, decision, and action, the OODA

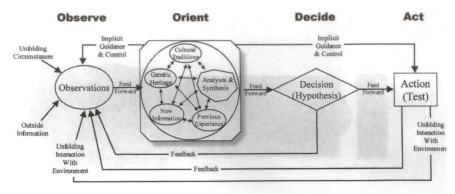

Figure drawing of Boyd's OODA Loop. (U.S. Air Force)

Loop. Boyd's influence reached far beyond the air force, contributing to a reshaping of Marine Corps doctrine and a broad spectrum of corporate practice.

Visibility and agility trumped speed and maneuverability. This insight suggested to Boyd a promising design approach that stressed the advantages of a lightweight and agile fighter aircraft armed with cheap, reliable weapons— 20-mm cannon and Sidewinder missiles. Such a machine would forgo high maximum speed for efficient cruising speed and carry an unusually large fraction of its weight in fuel for sustained combat. Small size combined with passive radar only would make the plane hard to detect either visually or electronically, an especially important consideration because combat experience since World War I proved that surprise was decisive in aerial combat. More than 80 percent of downed pilots learned they were under attack too late to do anything, or never knew what hit them at all.

Assigned to the Pentagon in 1966, Boyd had the opportunity to help turn theory into practice. Others, both in the Pentagon and elsewhere in the air force, had also been thinking about the problems of growing aircraft costs and declining combat effectiveness. Allied as the "Lightweight Fighter Mafia," they succeeded in convincing the Department of Defense to sponsor a competitive prototype development program, won in 1975 by General Dynamics with its YF-16. When F-16 Fighting Falcons reached air force units in 1979, it marked the first time since World War II that a new aircraft cost less and weighed less than its predecessor. Although it lacked the top speed and sophisticated avionics of the main air force air superiority McDonnell Douglas F-15 Eagle, the F-16 consistently outflew the F-15 in simulated combat.

PGM AND STEALTH

Precision-guided weapons and stealth technology joined information warfare as major shapers of thinking about the structure of armed forces in the late twentieth century. During its Vietnam debut, smart bombs had not impressed military users as anything more than a better way to put bombs on target. Only after the war did the idea that increasingly capable precision-guided munitions (PGM) represented something revolutionary, especially when stealth aircraft in the 1980s became available to deliver them. The junior officers of Vietnam thought a good deal about their experience as they climbed through the grades. When they commanded in 1991, they were ready to apply airpower in ways they were sure would be more effective.

Ultimately, the distinction between smart bomb and PGM ceased to matter much as their capabilities increasingly overlapped. By the time the Gulf War began, the more important distinction had become that between projectiles needing to be guided externally to their targets and those able to find their own targets after launching, often from considerable distance. The latter group now included cruise missiles, derived in part from the remotely piloted vehicles (RPVs) that had enjoyed a degree of success in Southeast Asia and the navy's medium-range Harpoon antiship missile deployed in the late 1970s. Their performance far outclassed that of the abortive cruise missiles of the 1950s. The best of the new generation, Tomahawk, carried terrain-matching radar so good that claims of pinpoint accuracy from hundreds of miles away seemed entirely plausible. Cruise missiles like Tomahawk could carry nuclear as well as conventional warheads. Other relatively short-range missiles shared that capability. Relatively low-yield nuclear warheads on such missiles raised the possibility, at least in theory, of so-called tactical nuclear warfare.

Although "stealth" has now become the shorthand term for a variety of techniques intended to render aircraft, missiles, and ships harder to detect by sight, sound, and heat as well as radar, the heart of the effort was defeating radar. To develop the required technology, the air force turned to Lockheed's Advanced Development Projects in California, the storied Skunk Works under its equally storied creator and director, Clarence "Kelly" Johnson. From the well-guarded windowless two-story concrete building next to the Burbank Airport had emerged America's first jet fighter (the P-80 Shooting Star), its first supersonic jet fighter (the F-104 Starfighter), and the U-2 spy plane. The first operational aircraft to include significant stealth features, the Lockheed SR-71 Blackbird, became operational in 1964. With a cruising speed of Mach 3 at 82,000 feet, the SR-71 could outfly any interceptor and most missiles, in the unlikely event of its detection, while continuously photographing all the territory it overflew.

Clarence Leonard "Kelly" Johnson (1910–1990)

Johnson was born in 1910, the son of Swedish immigrant parents who lived in a small Michigan mining town. Determined to become an airplane designer, he worked his way through the University of Michigan, which awarded him an M.S. in aeronautical engineering in

"Kelly" Johnson. (U.S. Air Force)

1933. He immediately went to work for Lockheed in California, and remained with the company for the rest of his life, rising to vice president for advanced projects development and board member. In 1937 he married Althea Louise Young, the first of his three wives. She died in 1969. In 1971 he married a young woman from New York named Mayellen Elberta Meade, who also died an early death. In 1980 he and Nancy Powers Horrigan married.

In the late 1930s, Johnson contributed to the unorthodox design of one of World War II's most successful fighters, Lockheed's twin-engine P-38 Lightning, the "fork-tailed devil" of the Pacific war. He also had a major hand in designing the much-used Lockheed Hudson bomber for the Royal Air Force. Meanwhile in 1938 Johnson moved to Lockheed's Burbank plant as chief research engineer. There he created in 1943 the design team for the first U.S. operational jet fighter, the Lockheed P-80 Shooting Star. The delivery of a proto-type in less than five months, thirty-seven days ahead of schedule, set the pattern for the remarkable R&D organization that came to be known as the Skunk Works, named after the highly secret Dogpatch manufacturer of Kickapoo Joy Juice in Al Capp's popular comic strip, *Li'l Abner*. Overall, Johnson designed or contributed to the design of over forty aircraft, most of them developed in secrecy at the Skunk Works.

As radar-guided missiles and gunfire grew ever more lethal over the next decade, the air force in 1978 contracted with Lockheed for a stealth attack plane. The Lockheed F-117A entered service in 1983 but its exis-tence remained a closely guarded secret until its spectacular appearance as a nearly undetectable platform for delivering PGM on Baghdad in the 1991 Gulf War. The Gulf War seemed to mark the apotheosis of American defense policy based on permanent technological revolution. High-tech warfare as displayed on television sets in the United States and around the world revealed the advanced stages of a third military-technological rev-olution in the making—after the classic military revolution of the sev-enteenth century, and the less widely acknowledged but perhaps even more decisive nineteenth-century revolution—but one the strained United States economy could no longer sustain without outside help.

Europe's twentieth century ended as it began, with war in the Balkans. But the century's military-technological transformations insured that the course of war would be quite different, once the major powers became involved. Memories of stalemated ground combat still haunted NATO commanders struggling over the role of airpower. "It was a struggle impelled by the losses in the trench warfare of World War I," observed Gen. Wesley K. Clark, Supreme Allied Commander Europe. "Everyone sought another, better way of fighting than that." In precision-guided weapons launched from aircraft, they may well have found the answer they sought, the critics notwithstanding. As events a few years later were to show, however, even the most sophisticated military technology could win battles, not wars.

PERMANENT TECHNOLOGICAL REVOLUTION

Trial and error had chiefly guided technological innovation for centuries, though applied science and systematic research began to achieve impressive results late in the nineteenth century. By the mid-twentieth century the balance had been reversed. Trial and error persisted, of course, but reasoned approaches to technological innovation could produce weapons to order. Just as engineering was becoming more systematic and mathematical, science was tending toward greater practicality. Military technology proliferated as applied science became ever less distinct from scientific engineering.

World War II reversed military attitudes toward research. Although innovative technologies played important and even vital roles in World War II, their source may have mattered even more than their actual contributions. Useful, even decisive, innovations flowing from wartime laboratories, both military and civilian, persuaded practical officers and officials that prewar visionaries had been right. Success so impressive, in sharp contrast to the futility of such efforts in the past, seemed to promise that directed innovation through applied research might become truly decisive in future war, that research and development would play major future roles in a new kind of war fought by machines instead of men. Not so clear was what kind of research, or who would be responsible for choosing among the alternatives.

Influence flowed both ways. War touched technologists and scientists as well as politicians and officers. Not only had they enjoyed virtually unlimited resources, a state of affairs many were eager to see sustained, they also found themselves attracted to military-funded research in unprecedented numbers. Military research by 1945 meant something quite different than it had before 1940. Perhaps more important, these were not the transient changes of the past. Institutions emerged after World War II to convert wartime arrangements into permanent features of American government and society. Permanent technological revolution became American military policy, even after it appeared to have met its match in Vietnam.

Unworkability plagued some conventional weapons systems, but their chief problems lay elsewhere. Preoccupation with technological sophistication led to weapons of enormous capabilities and equally enormous costs. Modern aircraft devoted half their weight to electronic gear, detectors and computers, collectively termed avionics. Avionics also accounted for much of the extraordinarily high price tag for such aircraft. Naval vessels have undergone similar evolution, and even the machines of ground combat have acquired sophisticated electronic capabilities and stunningly higher prices. Unrestrained pursuit of technical perfection has produced a baroque arsenal, possessed of extraordinary capabilities at astronomical expense.

Vietnam opened many eyes. Despite their undoubted technical superiority, the armed forces proved incapable of forcing a decision favorable to American war aims. Although they could dominate the battlefield, they could not win the war. The widespread assumption that better technology meant victory had made military-technological innovation, to a significant degree, an end in itself. During the decades after Vietnam, many officers, especially those who had seen frontline action, concluded that technological prowess alone was not enough and began thinking anew about the tactical and organizational ideas that might guide technological innovation into more useful channels.

Awesome victory in the Gulf War of 1991 seemed to validate the fresh thinking and rapidly changing technology that followed defeat in Southeast Asia. In the mid-1990s, despite ongoing ambiguities in the Balkans, Pentagon planners believed that America's armed forces verged on a new era, a "Revolution in Military Affairs" (RMA). Initially defined largely in technological terms, the phrase had, in fact, originated with Russian thinking a decade earlier about an imminent military-technological revolution based on radical increases in the ranges at which defenders could detect, engage, and defeat attacking armored forces without resorting to nuclear weapons. Although starting from the same appreciation for augmented battlefield reach, American theorists expanded the concept beyond the narrowly technical. Every aspect of military organization, training, and logistics would need reassessment in the light of new forms of combat.

The so-called revolution in part reflected the information-driven transformation of Western society in the late twentieth century. But to fast-paced advances in global communications networks and electronic data processing, the armed forces added precision-guided weapons and stealth technology, all displayed to extraordinary effect in the Gulf War of 1991. These newly merged capabilities appeared to require major changes in every aspect of military organization and activity from battlefield tactics to grand strategy. As the twentieth century came to a close, the rhetoric was shifting from revolution to transformation even as the relentless pace of technological change continued to accelerate.

Glossary

ABM. Antiballistic Missile System, a defense organized to destroy hostile ballistic missiles in flight.

Aircraft carrier. Naval ship capable of maintaining, launching, and recovering combat aircraft.

Airship. Self-propelled, steerable lighter-than-air flying machine.

American system. Method of manufacturing based on interchangeable parts, developed in U.S. Army arsenals during the early nineteenth century.

Armored fighting vehicle. Any self-propelled machine for use in combat.

Artillery. Crew-served guns.

Avionics. Aviation electronics, the electronic sensors and computers that make up half the weight of modern aircraft.

Ball. Spherical ammunition for smoothbore small arms.

Ballistic missile. Self-propelled projectile that follows a path solely determined by gravity (ballistic trajectory) after its fuel is expended.

Battleship. The largest warship in the post-sailing era naval line of battle.

Bayonet. Knife attached to the muzzle end of any shoulder arm, allowing it to be used as a kind of surrogate pike.

Bazooka. Tube-launched 2.36-inch fin-stabilized rocket fitted with an armor-piercing shaped charge, introduced in World War II.

Bore. Interior of a gun muzzle.

Breechloader. Any gunpowder weapon into which the projectile is loaded from the rear of the barrel into the firing chamber.

Bullet. Solid projectile fired from small arms and machine guns.

Caliber. Diameter of a projectile or of the bore of a gun or launching tube.

Camp follower. Person (often female) who accompanies an army in the field, providing services to the troops.

Caplock. Firing mechanism that uses the detonation of an unstable chemical compound to ignite the gunpowder.

Cavalry. Troops who fight mounted, formerly on horseback, more recently in motor vehicles.

Conventional weapons. Weapons that are not biological, chemical, or nuclear.

Convoy. Organized group of noncombatant vehicles or ships accompanied by armed defenders.

Cruise missile. Self-propelled pilotless aircraft with programmable flight path and self-contained terminal guidance.

Doctrine. Systematized principles for the use of armed forces or weapons systems.

Engineer. Troops who build and attack fortifications, and/or operate military machinery.

FBM. Fleet Ballistic Missile.

Fire control system. Mechanical or electronic components that direct the aim of a gun on its target.

Fission. Process of splitting heavy atoms to release energy, basis for the atomic bomb.

Flintlock. Firing mechanism that relies on sparks struck by flint against steel to ignite the gunpowder.

Fortification. Fixed or field works intended to shelter defenders against attackers.

Frigate. Sailing warship with single enclosed gun deck, carrying 24 to 44 guns.

Fusion. Process of uniting light elements to release energy, basis for H-bomb.

Guided missile. Projectile with wings, fins, or other aerodynamic surfaces that allow its flight path to be adjusted by external command or data from internal sensors.

Gun. Loosely, any device for firing a projectile with controlled explosion; more specifically, a firearm of larger caliber.

H-bomb. Thermonuclear or hydrogen bomb, an explosive device that uses the detonation of a fission bomb to ignite a fusion reaction.

HVAR. Five-inch high-velocity aircraft rocket for attacking ground targets, introduced in World War II.

ICBM. Intercontinental Ballistic Missile, ballistic missile with a range over 3,000 miles.

Infantry. Army combat forces that fight on foot.

Infiltration. Tactical method that bypasses defensive strong points for later attention in order to sustain the momentum of an attack.

IRBM. Intermediate-Range Ballistic Missile, ballistic missile with a range of 1,500–3,000 miles.

Jet plane. Aircraft propelled by a reaction engine that takes in oxygen from the atmosphere, burns it, and expels a high-speed exhaust to provide thrust.

Landing vehicle. Boat intended to land troops and equipment on hostile beaches.

Line. That part of a military organization with combat missions, usually the infantry, cavalry, and part of the artillery and engineers.

Lock. Firing mechanism of a firearm. *See also* **Caplock**; **Flintlock**; **Matchlock**.

Logistics. Science of moving and maintaining combat forces and their supplies.

Machine gun. Loosely, any small-caliber firearm, usually crew served, able to fire hundreds of projectiles a minute; more specifically, such a firearm that uses gas or recoil from firing to actuate loading and firing.

Matchlock. Firing mechanism using a smoldering wick (match) to ignite the gunpowder in a firearm.

Mechanization. Loosely, the equipment of military forces machines; more specifically, the process of equipping forces with armored fighting vehicles.

Militarism. Excessive admiration for the trappings of military life, activity, and organization.

MIRV. Multiple Independently targetable Reentry Vehicle (warhead).

Musket. Any smoothbore shoulder arm.

Muzzleloader. Any gunpowder weapon into which the projectile is inserted from the front end and pushed through the barrel to the firing chamber.

Navalism. Excessive admiration for the trappings of naval life, activity, and organization.

Operations. Direction of combat forces at a scale greater than tactics but lesser than strategy.

Operations analysis. Systematic and quantitative approaches to problems of military tactics and strategy.

PGM. Precision-guided munitions, bombs, or missiles that can be guided to their targets either by external control or by internal data processing.

Pike. Infantry weapon, an iron point mounted on a long wooden staff.

Quartermaster. Staff officer responsible for all supplies other than munitions.

Radar. Electronic device for detecting and determining the range and other characteristics of distant objects with radio waves.

Rifle. Any shoulder arm or artillery piece with rifled barrel.

Rifle-musket. U.S. Army model 1855 shoulder arm that added rifling to the barrels of standard caplock muskets.

Rifling. Physical structure of lands and grooves that imparts spin to a projectile as it traverses a gun barrel, increasing accuracy and range.

RMA. Revolution in Military Affairs.

Rocket. Reaction engine that carries its own fuel and oxidizer, making it independent of the atmosphere.

SAM. Surface-to-Air Missile, ground defense against attacking aircraft or missiles.

Satellite. Any orbiting body.

Shell. Hollow projectile filled with explosive.

Shoulder arm. Musket, rifle, or other relatively long individual firearm.

Siegecraft. Art and science of defending and attacking fortified places.

Small arm. Any firearm used by an individual.

Sonar. Electronic device for detecting and determining the range and other characteristics of distant underwater objects with sound waves.

Staff. Officers (and sometimes civilians) who assist a commander by attending to various support functions.

Strategic. Referring to that level of military activity that links national policy to military operations; also, more loosely, referring to military activity directed at the enemy's home front.

Strategy. Direction of armed forces at the highest level in implementing national policy.

Submarine. Loosely, any warship capable of launching attack while submerged; more specifically, a warship that operates mainly underwater.

Submersible. Warship able to briefly operate underwater.

Tactics. Military action on the battlefield.

Tank. Tracked armored fighting vehicle, usually with a turret-mounted large-caliber gun and several machine guns.

Thermonuclear. *See* **H-bomb**.

Triad. Trio of manned bombers, land-based missiles, and sea-based missiles organized to deter enemy nuclear attack.

Uniformity system. *See* **American system**.

V-1. German jet-propelled flying bomb (unmanned aircraft) introduced in 1944.

V-2. German rocket-propelled strategic missile introduced in 1944.

Volley. Simultaneous discharge of numerous firearms at a single target.

VT-fuze. Variable Time fuze, security designation of proximity fuze.

Zeppelin. German rigid airship used in World War I to bomb London.

Select Bibliography

GENERAL

Reference Works

Blair, Claude, ed. *Pollard's History of Firearms*. New York: Macmillan, 1983.

Chambers, John Whiteclay, ed. *The Oxford Companion to American Military History*. Oxford: Oxford University Press, 1999.

Garraty, John A., and Mark C. Carnes, eds. *American National Biography*. 24 vols. New York: Oxford University Press, 1999.

Higham, Robin, ed. *A Guide to the Sources of United States Military History*. Hamden, CT: Archon Books, 1975. Also *Supplement 1–IV*, ed. Higham and Donald J. Mrozek: 1981–1998.

Jessup, John E., Jr., and Louise B. Ketz, eds. *Encyclopedia of the American Military: Studies of the History, Traditions, Policies, Institutions, and Roles of the Armed Forces in War and Peace*. 3 vols. New York: Scribner's, 1994.

Kranzberg, Melvin, and Carroll L. Pursell, eds. *Technology in Western Civilization*. 2 vols. New York: Oxford University Press, 1967.

McHenry, Robert, ed. *Webster's American Military Biographies*. Springfield, MA: G. & C. Merriam, 1978.

Porter, Roy, and Marilyn Ogilvie, eds. *The Biographical Dictionary of Scientists*. 3d ed. 2 vols. New York: Oxford University Press, 2000.

Shrader, Charles R. *U.S. Military Logistics, 1607–1991: A Research Guide*. Research Guides to Military Studies 4. Westport, CT: Greenwood Press, 1992.

Shrader, Charles Reginald, ed. *Reference Guide to United States Military History.* 5 vols. New York: Facts on File, 1991–1995.

Spiller, Roger J., ed. *Dictionary of American Military Biography.* 3 vols. Westport, CT: Greenwood Press, 1984.

Who Was Who in American History—The Military. Chicago: Marquis Who's Who, 1975.

Who Was Who in American History—Science and Technology. Chicago: Marquis Who's Who, 1976.

Collections

Bradford, James C., ed. *Makers of the American Naval Tradition.* 3 vols. Annapolis, MD: Naval Institute Press, 1985–1990.

Gillis, John R., ed. *The Militarization of the Western World.* New Brunswick, NJ: Rutgers University Press, 1989.

Goldman, Emily O., and Leslie C. Eliason, eds. *The Diffusion of Military Technology and Ideas.* Stanford, CA: Stanford University Press, 2003.

Hacker, Barton C., and Margaret Vining, eds. *Science in Uniform, Uniforms in Science: Historical Studies of American Military and Scientific Interactions.* Lanham, MD: Scarecrow Press, in press.

Hagan, Kenneth J., ed. *In Peace and War: Interpretations of American Naval History, 1775–1978.* Westport, CT: Greenwood Press, 1978.

Hagan, Kenneth J., and William R. Roberts, eds. *Against All Enemies: Interpretations of American Military History from Colonial Times to the Present.* New York: Greenwood Press, 1986.

Holmes, Richard, ed. *The World Atlas of Warfare: Military Innovations That Changed the Course of History.* New York: Viking Studio Books, 1988.

King, Randolph W., Prescott Palmer, and Bruce I. Meader, eds. *Naval Engineering and American Seapower.* Baltimore: Nautical & Aviation Publishing Company of America, 1989.

Knox, MacGregor, and Williamson Murray, eds. *The Dynamics of Military Revolution, 1300–2050.* Cambridge: Cambridge University Press, 2001.

Morison, Elting E. *Men, Machines, and Modern Times.* Cambridge, MA: MIT Press, 1966.

Paret, Peter, ed. *Makers of Modern Strategy: From Machiavelli to the Nuclear Age.* Princeton, NJ: Princeton University Press, 1986.

Parker, Geoffrey, ed. *The Cambridge Illustrated History of Warfare: The Triumph of the West.* Cambridge: Cambridge University Press, 1995.

Pursell, Carroll W., Jr., ed. *Technology in America: A History of Individuals and Ideas,* 2d ed. Cambridge, MA: MIT Press, 1990.

Reingold, Nathan. *Science, American Style.* New Brunswick, NJ: Rutgers University Press, 1991.

Reingold, Nathan, ed. *The Sciences in the American Context: New Perspectives.* Washington, DC: Smithsonian Institution Press, 1979.

Smith, Merritt Roe, ed. *Military Enterprise and Technological Change: Perspectives on the American Experience*. Cambridge, MA: MIT Press, 1984.

Townshend, Charles, ed. *The Oxford Illustrated History of Modern War*. New York: Oxford University Press, 1997.

Books

Alger, John I. *The Quest for Victory: The History of the Principles of War*. Westport, CT: Greenwood Press, 1982.

Basalla, George. *The Evolution of Technology*. Cambridge: Cambridge University Press, 1988.

Crackel, Theodore J. *West Point: A Bicentennial History*. Lawrence: University Press of Kansas, 2002.

Dupree, A. Hunter. *Science in the Federal Government: A History of Policies and Activities to 1940*. Cambridge, MA: Harvard University Press, 1957.

Ellis, John. *The Social History of the Machine Gun*. New York: Pantheon, 1975.

Gillett, Mary C. *The Army Medical Department*. 3 vols. Washington, DC: Center of Military History, 1981–1995.

Hughes, Thomas P. *American Genesis: A Century of Invention and Technological Enthusiasm, 1870–1970*. New York: Viking, 1989.

Hunter, Louis C. *A History of Industrial Power in the United States, 1780–1930*. 3 vols. Charlottesville: University Press of Virginia, 1979–1991.

Huntington, Samuel P. *The Soldier and the State: The Theory and Politics of Civil-Military Relations*. Cambridge, MA: Harvard University Press, 1959.

Kaufmann, Joseph E., and H. W. Kaufmann. *Fortress America: The Forts That Defended America, 1600 to the Present*. New York: Da Capo Press, 2004.

Koistinen, Paul A. C. *The Political Economy of American Warfare*. 4 vols. Lawrence: University Press of Kansas, 1996–2004.

McBride, William M. *The Rise and Fall of a Strategic Technology: The American Battleship from Santiago Bay to Pearl Harbor, 1898–1941*. Baltimore: Johns Hopkins University Press, 1990.

———. *Technological Change and the United States Navy, 1895–1945*. Baltimore: Johns Hopkins University Press, 2000.

McNeill, William H. *The Pursuit of Power: Technology, Armed Force, and Society since A.D. 1000*. Chicago: University of Chicago Press, 1983.

Millett, Alan R. *Semper Fidelis: The History of the United States Marine Corps*. New York: Macmillan, 1980.

Millis, Walter. *Arms and Men: A Study of American Military History*. New York: G. P. Putnam's Sons, 1956.

Nenninger, Timothy K. *The Leavenworth Schools and the Old Army: Education, Professionalism, and the Officer Corps of the United States Army, 1881–1918*. Westport, CT: Greenwood Press, 1978.

Raines, Rebecca Robbins. *Getting the Message Through: A Branch History of the U.S. Army Signal Corps*. Army Historical Series. Washington, DC: Center of Military History, 1996.

Skowronek, Stephen. *Building a New American State: The Expansion of National Administrative Capacities, 1877–1920*. Cambridge: Cambridge University Press, 1982.

Sweetman, Jack. *The U.S. Naval Academy: An Illustrated History*. Annapolis, MD: Naval Institute Press, 1979.

Weigley, Russell F. *The American Way of War: A History of United States Military Strategy and Policy*. New York: Macmillan, 1973.

WORKS DEALING PRIMARILY WITH THE NINETEENTH CENTURY AND EARLIER (CHAPTERS 1–4)

Reference Works

Chapelle, Howard I. *The History of the American Sailing Navy: The Ships and Their Development*. New York: W. W. Norton, 1949.

Coggins, Jack. *Arms and Equipment of the Civil War*. Garden City, NY: Doubleday, 1962.

———. *Ships and Seamen of the American Revolution: Vessels, Crews, Weapons, Gear, Naval Tactics, and Actions of the War for Independence*. Harrisburg, PA: Stackpole Books, 1969.

Edwards, William B. *Civil War Guns. The Complete Story of Federal and Confederate Small Arms: Design, Manufacture, Identification, Procurement, Issue Employment, Effectiveness, and Postwar Disposal*. Harrisburg, PA: Stackpole Books, 1962.

Moller, George D. *American Military Shoulder Arms*, vol. 1, *Colonial and Revolutionary War Arms*. Niwot: University Press of Colorado, 1993.

Myatt, Frederick. *The Illustrated Encyclopedia of 19th Century Firearms: An Illustrated History of the Development of the World's Military Firearms during the 19th Century*. New York: Crescent Books, 1979.

Peterson, Harold L. *The Book of the Continental Soldier: Being a Compleat Account of the Uniforms, Weapons, and Equipment with Which He Lived and Fought*. Harrisburg, PA: Stackpole Books, 1968.

Ripley, Warren. *Artillery and Ammunition of the Civil War*. New York: Van Nostrand Reinhold, 1970.

Silverstone, Paul H. *Warships of the Civil War Navies*. Annapolis, MD: Naval Institute Press, 1989.

Time-Life Books. *Echoes of Glory: Arms and Equipment of the Confederacy*. Alexandria, VA: Time-Life Books, 1991

———. *Echoes of Glory: Arms and Equipment of the Union*. Alexandria, VA: Time-Life Books, 1991.

Collections

Förster, Stig, and Jörg Nagler, eds. *On the Road to Total War: The American Civil War and the German Wars of Unification, 1861–1871.* Washington, DC: German Historical Institute; Cambridge: Cambridge University Press, 1997.

Gardiner, Robert, ed. *Steam, Steel & Shellfire: The Steam Warship, 1815–1905.* Annapolis, MD: Naval Institute Press, 1992.

Mayr, Otto, and Robert C. Post, eds. *Yankee Enterprise: The Rise of the American System of Manufacturing.* Washington, DC: Smithsonian Institution Press, 1981.

Books

Alden, John D. *The American Steel Navy.* Annapolis, MD: Naval Institute Press, 1972.

Angevine, Robert G. *The Railroad and the State: War, Politics, and Technology in Nineteenth-Century America.* Stanford, CA: Stanford University Press, 2004.

Armstrong, David A. *Bullets and Bureaucrats: The Machine Gun and the United States Army, 1861–1916.* Westport, CT: Greenwood Press, 1982.

Bacon, Benjamin W. *Sinews of War: How Technology, Industry, and Transportation Won the Civil War.* Novato, CA: Presidio, 1997.

Brown, M. L. *Firearms in Colonial America: The Impact on History and Technology, 1492–1792.* Washington, DC: Smithsonian Institution Press, 1980.

Bruce, Robert V. *The Launching of Modern American Science, 1846–1876.* New York: Alfred A. Knopf, 1987.

Cirillo, Vincent J. *Bullets and Bacilli: The Spanish-American War and Military Medicine.* New Brunswick, NJ: Rutgers University Press, 2004.

Davis, Carl L. *Arming the Union: Small Arms in the Union Army.* Port Washington, NY: Kennikat Press, 1973.

Delaporte, François. *The History of Yellow Fever: An Essay on the Birth of Tropical Medicine.* Translated by Arthur Goldhammer. Cambridge, MA: MIT Press, 1991.

Epstein, Robert M. *Napoleon's Last Victory and the Emergence of Modern War.* Lawrence: University Press of Kansas, 1994.

Goetzmann, William H. *Army Exploration in the American West, 1803–1863.* New Haven, CT: Yale University Press, 1959.

———. *Exploration and Empire: The Explorer and the Scientist in the Winning of the American West.* New York: Alfred A. Knopf, 1966.

———. *New Lands, New Men: America and the Second Great Age of Discovery.* New York: Viking Penguin, 1986.

Gray, Edwyn. *The Devil's Device: Robert Whitehead and the History of the Torpedo.* Rev. ed. Annapolis, MD: Naval Institute Press, 1991.

Griffith, Paddy. *Battle Tactics of the Civil War.* New Haven, CT: Yale University Press, 1989.

Hackemer, Kurt. *The U.S. Navy and the Origins of the Military-Industrial Complex, 1847–1883*. Annapolis, MD: Naval Institute Press, 2001.

Hagerman, Edward. *The American Civil War and the Origins of Modern Warfare*. Bloomington: Indiana University Press, 1988.

Hampton, H. Duane. *How the U.S. Cavalry Saved Our National Parks*. Bloomington: Indiana University Press, 1971.

Hess, Earl J. *Field Armies and Fortifications in the Civil War: The Eastern Campaigns, 1861–1864*. Chapel Hill: University of North Carolina Press, 2005.

Hill, Forest G. *Roads, Rails and Waterways: The Army Engineers and Early Transportation*. Norman: University of Oklahoma Press, 1957.

Hindle, Brooke. *The Pursuit of Science in Revolutionary America, 1735–1789*. Chapel Hill: University of North Carolina Press, 1956.

Karsen, Peter. *The Naval Aristocracy: The Golden Age of Annapolis and the Emergence of Modern American Navalism*. Riverside, NJ: Free Press, 1972.

Lash, Jeffrey N. *Destroyer of the Iron Horse: Joseph E. Johnston and Confederate Rail Transport, 1861–1865*. Kent, OH: Kent State University Press, 1991.

Meredith, Roy, and Arthur Meredith. *Mr. Lincoln's Military Railroads: A Pictorial History of the U.S. Civil War Railroads*. New York: W. W. Norton, 1979.

Mindell, David A. *War, Technology, and Experience aboard the USS Monitor*. Baltimore: Johns Hopkins University Press, 2000.

O'Connell, Robert L. *Sacred Vessels: The Cult of the Battleship and the Rise of the U.S. Navy*. New York: Oxford University Press, 1991.

Pappas, George S. *To the Point: The United States Military Academy, 1802–1902*. New York: Praeger, 1993.

Ponko, Vincent, Jr. *Ships, Seas, and Scientists: U.S. Naval Exploration and Discovery in the Nineteenth Century*. Annapolis, MD: Naval Institute Press, 1974.

Roberts, William H. *Civil War Ironclads: The U.S. Navy and Industrial Mobilization*. Baltimore: Johns Hopkins University Press, 2002.

Roland, Alex. *Underwater Warfare in the Age of Sail*. Bloomington: Indiana University Press, 1978.

Sale, Kirkpatrick. *The Fire of His Genius: Robert Fulton and the American Dream*. New York: Free Press, 2001.

Schubert, Frank N. *Vanguard of Expansion: Army Engineers in the Trans-Mississippi West, 1819–1879*. Washington, DC: Office of the Chief of Engineers, 1980.

Shallat, Todd. *Structures in the Stream: Water, Science, and the Rise of the U.S. Army Corps of Engineers*. Austin: University of Texas Press, 1994.

Sherwood, Morgan B. *Exploration of Alaska, 1865–1900*. New Haven, CT: Yale University Press, 1965.

Shulman, Mark Russell. *Navalism and the Emergence of American Sea Power, 1882–1893*. Annapolis, MD: Naval Institute Press, 1995.

Smith, Merritt Roe. *Harpers Ferry Armory and the New Technology: The Challenge of Change*. Ithaca, NY: Cornell University Press, 1977.

Sumida, Jon Tetsuro. *Inventing Grand Strategy and Teaching Command: The Classic Works of Alfred Thayer Mahan Reconsidered*. Baltimore: Johns Hopkins University Press, 1997.

York, Neil Longley. *Mechanical Metamorphosis: Technological Change in Revolutionary America*. Westport, CT: Greenwood Press, 1985.

WORKS DEALING PRIMARILY WITH THE TWENTIETH CENTURY (CHAPTERS 5–11)

Reference Works

Angelucci, Enzo. *Rand McNally Encyclopedia of Military Aircraft, 1914 to the Present*. Translated by S. M. Harris. Chicago: Rand McNally, 1981.

Frankland, Noble, ed. *The Encyclopedia of Twentieth Century Warfare*. New York: Crown, 1989.

Friedman, Norman. *Submarine Design and Development*. Annapolis, MD: Naval Institute Press, 1984.

———. *U.S. Aircraft Carriers: An Illustrated Design History*. Annapolis, MD: Naval Institute Press, 1983.

———. *U.S. Naval Weapons: Every Gun, Missile, Mine, and Torpedo Used by the U.S. Navy from 1883 to the Present Day*. Annapolis, MD: Naval Institute Press, 1982.

Gunston, Bill. *The Illustrated Encyclopedia of the World's Rockets and Missiles: A Comprehensive Technical Directory and History of the Military Guided Missile Systems of the 20th Century*. New York: Crescent Books, 1979.

Moore, John, ed. *Jane's American Fighting Ships of the 20th Century*. New York: Mallard Press, 1991.

Swanborough, Gordon, and Peter M. Bowers. *United States Navy Aircraft since 1911*. 3d ed. Annapolis, MD: Naval Institute Press, 1990.

Taylor, Michael J. H. *Jane's American Fighting Aircraft of the 20th Century*. New York: Mallard Press, 1991.

Collections

Forman, Paul, and José M. Sánchez-Ron, eds. *National Military Establishments and the Advancement of Science and Technology*. Dordrecht, the Netherlands: Kluwer Academic, 1996.

Galison, Peter, and Bruce Hevly, eds. *Big Science: The Growth of Large-Scale Research*. Stanford, CA: Stanford University Press, 1992.

Harris, J. P., and F. H. Toase, eds. *Armoured Warfare*. London: Batsford, 1990.

Holton, Gerald, ed. *The Twentieth-Century Sciences: Studies in the Biography of Ideas*. New York: W. W. Norton, 1972.

Hughes, Agatha C., and Thomas P. Hughes, eds. *Systems, Experts, and Computers: The Systems Approach in Management and Engineering, World War II and After*. Cambridge, MA: MIT Press, 2000.

Krige, John, and Dominique Pestre, eds. *Science in the Twentieth Century*. Amsterdam: Harwood Academic, 1997.

McInnes, Colin, and G. D. Sheffield, eds. *Warfare in the Twentieth Century: Theory and Practice*. London: Unwin Hyman, 1988.

Meilinger, Phillip S., ed. *The Paths of Heaven: The Evolution of Airpower Theory*. Maxwell Air Force Base, Montgomery, AL: Air University Press, 1997.

Mendelsohn, Everett, Merritt Roe Smith, and Peter Weingart, eds. *Science, Technology and the Military*. 2 vols. Dordrecht, the Netherlands: Kluwer Academic, 1988.

Neufeld, Jacob, George M. Watson, Jr., and David Chenoweth, eds. *Technology and the Air Force: A Retrospective Assessment*. Washington, DC: Air Force History and Museums Program, 1997.

Schwartz, Stephen I., ed. *Atomic Audit: The Costs and Consequences of U.S. Nuclear Weapons since 1940*. Washington, DC: Brookings Institution, 1998.

Books

Bruce-Briggs, B. *The Shield of Faith: A Chronicle of Strategic Defense from Zeppelins to Star Wars*. New York: Simon & Schuster, 1988.

Ceruzzi, Paul E. *A History of Modern Computing*. Cambridge, MA: MIT Press, 1998.

Franklin, H. Bruce. *War Stars: The Superweapon and the American Imagination*. New York: Oxford University Press, 1988.

Hart, David M. *Forged Consensus: Science, Technology, and Economic Policy in the United States, 1921–1953*. Princeton, NJ: Princeton University Press, 1998.

Hewlett, Richard G., Oscar E. Anderson, Francis Duncan, and Jack M. Holl. *A History of the Atomic Energy Commission*, 3 vols. University Park: Pennsylvania State University Press; 1962–1969; Berkeley: University of California Press, 1989.

Kaldor, Mary. *The Baroque Arsenal*. New York: Hill & Wang, 1981.

Kennett, Lee. *A History of Strategic Bombing*. New York: Scribner's, 1982.

Mason, Herbert Molloy, Jr. *The United States Air Force: A Turbulent History*. New York: Mason/Charter, 1976.

Mindell, David A. *Between Human and Machine: Feedback, Control, and Computing before Cybernetics*. Baltimore: Johns Hopkins University Press, 2002.

Mowery, David C., and Nathan Rosenberg. *Paths of Innovation: Technological Change in 20th Century America*. New York: Cambridge University Press, 1998.

Noble, David F. *Forces of Production: A Social History of Industrial Automation*. New York: Alfred A. Knopf, 1984.

Ogorkiewicz, Richard M. *Armoured Forces: A History of Armoured Forces and Their Vehicles*. 2d ed. New York: Arco, 1970.

Roland, Alex. *Model Research: The National Advisory Committee for Aeronautics, 1915–1958*. 2 vols. Washington, DC: National Aeronautics and Space Administration, 1985.

Russell, Edmund. *War and Nature: Fighting Humans and Insects with Chemicals from World War I to "Silent Spring."* Cambridge: Cambridge University Press, 2001.

Smith, Peter C. *Close Air Support: An Illustrated History, 1914 to the Present*. New York: Orion Books, 1990.

Weir, Gary E. *Forged in War: The Naval-Industrial Complex and American Submarine Construction, 1940–1961*. Washington, DC: Naval Historical Center, 1993.

WORKS DEALING PRIMARILY WITH WORLD WARS I AND II (CHAPTERS 5–7)

Reference Works

Ellis, Chris, and Denis Bishop. *Military Transport of World War I, Including Vintage Vehicles and Post War Models*. New York: Macmillan, 1970.

Friedman, Norman. *U.S. Submarines through 1945: An Illustrated Design History*. Annapolis, MD: Naval Institute Press, 1995.

Luttwak, Edward N., and Stuart L. Koehl. *The Dictionary of Modern War*. New York: HarperCollins, 1991.

Tucker, Spencer, ed. *The European Powers in the First World War: An Encyclopedia*. New York: Garland, 1999.

Venzon, Anne Cipriano, ed. *The United States in the First World War: An Encyclopedia*. New York: Garland, 1995.

White, B. T. *Tanks and Other Armored Fighting Vehicles, 1900 to 1918*. Mechanised Warfare in Color. New York: Macmillan, 1970.

Collections

Chickering, Roger, and Stig Förster, eds. *Great War, Total War: Combat and Mobilization on the Western Front, 1914–1918*. Washington, DC: German Historical Institute; Cambridge: Cambridge University Press, 2000.

Dreisziger, Nandor Fred, ed. *Mobilization for Total War: The Canadian, American and British Experience, 1914–1918, 1939–1945*. Waterloo, Ontario: Wilfrid Laurier University Press, 1981.

Miller, Steven E., ed. *Military Strategy and the Origins of the First World War*. Princeton, NJ: Princeton University Press, 1985.

Murray, Williamson, and Allan R. Millett, eds. *Military Innovation in the Interwar Period*. Cambridge: Cambridge University Press, 1996.

Strachan, Hew, ed. *World War I: A History*. New York: Oxford University Press, 1998.

Books

Baldwin, Ralph. *The Deadly Fuze: The Secret Weapon of World War II*. San Rafael, CA: Presidio Press, 1980.

Baxter, James Phinney, III. *Scientists against Time*. History of the Office of Scientific Research and Development: Science in World War II. Boston: Little, Brown, 1946.

Belote, James H., and William M. Belote. *Titans of the Sea: The Development and Operations of Japanese and American Carrier Task Forces during World War II*. New York: Harper & Row, 1975.

Benedict, Ruth. *The Chrysanthemum and the Sword: Patterns of Japanese Culture*. Boston: Houghton Mifflin, 1946.

Biddle, Tami Davis. *Rhetoric and Reality in Air Warfare: The Evolution of British and American Ideas about Strategic Bombing, 1914–1945*. Princeton, NJ: Princeton University Press, 2002.

Bidwell, Shelford, and Dominick Graham. *Fire-Power: British Army Weapons and Theories of War, 1904–1945*. London: George Allen & Unwin, 1982.

Brown, Louis. *A Radar History of World War II: Technical and Military Imperatives*. Bristol, CT and Philadelphia, PA: Institute of Physics Publishing, 1999.

Byerly, Carol R. *Fever of War: The Influenza Epidemic in the U.S. Army during World War I*. New York: New York University Press, 2005.

Committee for the Compilation of Materials on Damage Caused by the Atomic Bombs in Hiroshima and Nagasaki. *Hiroshima and Nagasaki: The Physical, Medical, and Social Effects of the Atomic Bombing*. Translated by Eisei Ishikawa and David L. Swain. New York: Basic Books, 1981.

Cowdrey, Albert E. *Fighting for Life: American Military Medicine in World War II*. New York: Free Press, 1994.

Ellis, John. *Eye-Deep in Hell: Trench Warfare in World War I*. New York: Pantheon Books, 1976.

Faltum, Andrew. *The Essex Aircraft Carriers*. Baltimore: Nautical & Aviation, 1996.

Ferguson, Niall. *The Pity of War*. New York: Basic Books, 1999.

Fletcher, David, ed. *Tanks and Trenches: First Hand Accounts of Tank Warfare in the First World War*. Stroud, England: Sutton, 1994.

Haber, Lutz F. *The Poisonous Cloud: Chemical Warfare in the First World War*. Oxford: Clarendon Press, 1986.

Hallion, Richard P. *Rise of the Fighter Aircraft, 1914–1918*. Annapolis, MD: Nautical & Aviation, 1984.

Hansen, Arlen J. *Gentlemen Volunteers: The Story of American Ambulance Drivers in the Great War, August 1914–September 1918*. New York: Arcade, 1996.

Hartcup, Guy. *The Effect of Science on the Second World War*. New York: St. Martin's Press, 2000.

———. *The War of Invention: Scientific Developments, 1914–1918*. London: Brassey's, 1988.

Hoddeson, Lillian, Paul W. Henriksen, Roger A. Meade, et al. *Critical Assembly: A Technical History of Los Alamos during the Oppenheimer Years, 1943–1945*. New York: Cambridge University Press, 1993.

Holley, Irving Brinton, Jr. *Ideas and Weapons: Exploitation of the Aerial Weapon by the United States during World War I; A Study in the Relationship of*

Technological Advance, Military Doctrine, and the Development of Weapons. New Haven, CT: Yale University Press, 1953.

Iseley, Jeter A., and Philip A. Crowl. *The U.S. Marines and Amphibious War: Its Theory, and Its Practice in the Pacific.* Princeton, NJ: Princeton University Press, 1951.

Johnson, David E. *Fast Tanks and Heavy Bombers: Innovation in the U.S. Army, 1917–1945.* Ithaca, NY: Cornell University Press, 1998.

Johnson, Hubert C. *Breakthrough! Tactics, Technology, and the Search for Victory on the Western Front in World War I.* Novato, CA: Presidio Press, 1994.

Jones, Neville. *The Origins of Strategic Bombing: A Study of the Development of British Air Strategic Thought and Practice up to 1918.* London: William Kimber, 1973.

Jones, Vincent C. *Manhattan: The Army and the Atomic Bomb.* Washington, DC: Center of Military History, 1985.

Kelsey, Benjamin S. *The Dragon's Teeth? The Creation of United States Air Power for World War II.* Washington, DC: Smithsonian Institution Press, 1982.

Kennett, Lee B. *The First Air War, 1914–1918.* New York: Free Press, 1991.

Leed, Eric J. *No Man's Land: Combat and Identity in World War I.* Cambridge: Cambridge University Press, 1979.

Levine, Alan J. *The Strategic Bombing of Germany, 1940–1945.* New York: Praeger, 1992.

Meigs, Montgomery C. *Slide Rules and Submarines: American Scientists and Sub-surface Warfare in World War II.* Washington, DC: National Defense University Press, 1990.

Mierzejewski, Alfred C. *The Collapse of the German War Economy, 1944–1945: Allied Air Power and the German National Railway.* Chapel Hill: University of North Carolina Press, 1988.

Morrow, John H., Jr. *The Great War in the Air: Military Aviation from 1909 to 1921.* Washington, DC: Smithsonian Institution Press, 1993.

Moy, Timothy. *War Machines: Transforming Technologies in the U.S. Military, 1920–1940.* College Station: Texas A&M University Press, 2001.

Neufeld, Michael J. *The Rocket and the Reich: Peenemünde and the Coming of the Ballistic Missile Era.* New York: Free Press, 1995.

Odom, William O. *After the Trenches: The Transformation of U.S. Military Doctrine, 1918–1939.* Texas A&M Military History Series 64. College Station: Texas A&M University Press, 1999.

Overy, Richard. *Why the Allies Won.* New York: W. W. Norton, 1995.

Rhodes, Richard. *The Making of the Atomic Bomb.* New York: Simon & Schuster, 1986.

Robinson, Douglas H. *The Zeppelins in Combat: A History of the German Naval Airship Division, 1912–1918.* 3d ed. London: G. T. Foulis, 1971.

Ross, Stewart Halsey. *Strategic Bombing by the United States in World War II: The Myths and the Facts.* Jefferson, NC: McFarland, 2003.

Schaffer, Ronald. *Wings of Judgment: American Bombing in World War II.* New York: Oxford University Press, 1985.

Shachtman, Tom. *Laboratory Warriors: How Allied Science and Technology Tipped the Balance in World War II.* New York: William Morrow, 2000.

Sherry, Michael S. *The Rise of American Air Power: The Creation of Armageddon.* New Haven, CT: Yale University Press, 1987.

Snyder, Jack. *The Ideology of the Offensive: Military Decision Making and the Disasters of 1914.* Ithaca, NY: Cornell University Press, 1984.

Spector, Ronald H. *Eagle against the Sun: The American War with Japan.* New York: Free Press, 1985.

Strahan, Jerry E. *Andrew Jackson Higgins and the Boats That Won World War II.* Baton Rouge: Louisiana State University Press, 1994.

Syrett, David. *The Defeat of the German U-Boats: The Battle of the Atlantic.* Columbia: University of South Carolina Press, 1994.

Tarrant, V. E. *The U-boat Offensive, 1914–1945.* Annapolis, MD: Naval Institute Press, 1989.

Terraine, John. *The U-Boat Wars, 1916–1945.* New York: Putnam's, 1989.

Travers, Tim. *The Killing Ground: The British Army, the Western Front, and the Emergence of Modern Warfare, 1900–1918.* London: Allen & Unwin, 1987.

Waddington, C. H. *OR in World War 2: Operational Research against the U-Boat.* London: Elek Science, 1972.

Werrell, Kenneth P. *Blankets of Fire: U.S. Bombers over Japan during World War II.* Washington, DC: Smithsonian Institution Press, 1996.

Williams, Kathleen Broome. *Secret Weapon: U.S. High-Frequency Direction Finding in the Battle of the Atlantic.* Annapolis, MD: Naval Institute Press, 1996.

Wilson, Dale E. *Treat 'Em Rough: The Birth of American Armor, 1917–1920.* Novato, CA: Presidio Press, 1989.

Winter, Jay, Geoffrey Parker, and Mary R. Habeck, eds. *The Great War and the Twentieth Century.* New Haven, CT: Yale University Press, 2000.

Woodman, Harry. *Early Aircraft Armament: The Aeroplane and the Gun up to 1918.* Washington, DC: Smithsonian Institution Press, 1989.

Zabecki, David T. *Steel Wind: Colonel Georg Bruchmüller and the Birth of Modern Artillery.* Westport, CT: Praeger, 1994.

Zimmerman, David. *Top Secret Exchange: The Tizard Mission and the Scientific War.* Stroud, England: Sutton; Montreal, Canada: McGill-Queen's University, 1996.

WORKS DEALING PRIMARILY WITH THE LATER TWENTIETH CENTURY (CHAPTERS 8–11)

Reference Works

Clark, Gregory R. *The Slang, Jargon, Abbreviations, Acronyms, Nomenclature, Nicknames, Pseudonyms, Slogans, Specs, Euphemisms, Double-talk, Chants, and*

Names and Places of the Era of United States Involvement in Vietnam. Jefferson, SC: McFarland, 1990.

Friedman, Norman. *U.S. Submarines since 1945: An Illustrated Design History*. Annapolis, MD: Naval Institute Press, 1994.

Hansen, Chuck. *U.S. Nuclear Weapons: The Secret History*. Arlington, TX: Aerofax, 1988.

Lawrence, Robert M. *Strategic Defense Initiative: Bibliography and Research Guide*. Boulder, CO: Westview Press, 1987.

Levine, Alan J. *The Missile and Space Race*. Westport, CT: Praeger, 1994.

Waller, Douglas C., James T. Bruce, Jr., and Douglas M. Cook. *The Strategic Defense Initiative: Progress and Challenges. A Guide to the Issues and References*. Claremont, CA: Regina Books, 1987.

Collections

Bellin, David, and Gary Chapman, eds. *Computers in Battle: Will They Work?* Boston: Harcourt Brace Jovanovich, 1987.

Brahmstedt, Christian, ed. *Defense's Nuclear Agency, 1947–1997*. Washington, DC: Defense Threat Reduction Agency, 2002.

Day, Dwayne A., John M. Logsdon, and Brian Latell, eds. *Eye in the Sky: The Story of the Corona Spy Satellites*. Washington, DC: Smithsonian Institution Press, 1998.

Gongora, Thierry, and Harald von Riekhoff, eds. *Toward a Revolution in Military Affairs? Defense and Security at the Dawn of the Twenty-First Century*. Westport, CT: Greenwood Press, 2000.

Horowitz, Irving Louis, ed. *The Rise and Fall of Project Camelot: Studies in the Relationship between Social Sciences and Practical Politics*. Cambridge, MA: MIT Press, 1967.

Jensen, Geoffrey, and Andrew Wiest, eds. *War in the Age of Technology: Myriad Faces of Modern Conflict*. New York: New York University Press, 2001.

Kemp, Geoffrey, Robert L. Pfaltzgraff, Jr., and Uri Ra'anan, eds. *The Other Arms Race: New Technologies and Non-Nuclear Conflict*. Lexington, MA: Lexington Books, 1975.

Schiffrin, André, ed. *The Cold War and the University: Toward an Intellectual History of the Postwar Years*, New York: New Press, 1997.

Tirman, John, ed. *The Militarization of High Technology*. Cambridge, MA: Ballinger, 1984.

Tucker, Samuel A., ed. *National Security Management: A Modern Design for Defense Decision. A McNamara-Hitch-Enthoven Anthology*. Washington, DC: Industrial College of the Armed Forces, 1966.

Wrage, Stephen D., ed. *Immaculate Warfare: Participants Reflect on the Air Campaigns over Kosovo, Afghanistan, and Iraq*. Westport, CT: Praeger, 2003.

Books

Abbate, Janet. *Inventing the Internet*. Cambridge, MA: MIT Press, 1999.

Allen, Matthew. *Military Helicopter Doctrines of the Major Powers, 1945–1992*. Westport, CT: Greenwood Press, 1993.

Arnold, David Christopher. *Spying from Space: Constructing America's Satellite Command and Control Systems*. College Station: Texas A&M University Press, 2005.

Ballard, Jack. *The Development and Employment of Fixed-Wing Gunships, 1962–1972*. Washington, DC: Office of Air Force History, 1982.

Barnaby, Frank. *The Automated Battlefield*. New York: Free Press, 1986.

Baucom, Donald R. *The Origins of SDI, 1944–1983*. Lawrence: University Press of Kansas, 1992.

Beard, Edmund. *Developing the ICBM: A Study in Bureaucratic Politics*. New York: Columbia University Press, 1976.

Bergen, John D. *Military Communications: A Test for Technology*. United States Army in Vietnam. Washington, DC: Center of Military History, 1986.

Boyne, Walter J. *Boeing B-52: A Documentary History*. Washington, DC: Smithsonian Institution Press, 1981.

Bruno, Lester H., and Richard Dean Burns. *The Quest for Missile Defenses, 1944–2003*. Claremont, CA: Regina Books, 2003.

Butrica, Andrew J. *Beyond the Ionosphere: Fifty Years of Satellite Communication*. Washington, DC: National Aeronautics and Space Administration, 1997.

Clark, Wesley K. *Waging Modern War: Bosnia, Kosovo, and the Future of Combat*. New York: Public Affairs, 2001.

Comor, Edward A. *Communication, Commerce, and Power: The Political Economy of America and the Direct Broadcast Satellite*. New York: St. Martin's Press, 1998.

Coram, Robert. *Boyd: The Fighter Pilot Who Changed the Art of War*. Boston: Little, Brown, 2002.

DeVorkin, David H. *Science with a Vengeance: How the Military Created the U.S. Space Sciences after World War II*. New York: Springer Verlag, 1992.

Dickson, Paul. *The Electronic Battlefield*. Bloomington: Indiana University Press, 1976.

Divine, Robert A. *The Sputnik Challenge*. New York: Oxford University Press, 1993.

Dorland, Peter, and James Nanney. *Dust Off: Army Aeromedical Evacuation in Vietnam*. Washington, DC: Center of Military History, 1982.

Dunnigan, James F., and Austin Bay. *From Shield to Storm: High-Tech Weapons, Military Strategy, and Coalition Warfare in the Persian Gulf*. New York: William Morrow, 1992.

Edwards, Paul N. *The Closed World: Computers and Politics in Cold War America*. Cambridge, MA: MIT Press, 1996.

Evangelista, Matthew. *Innovation and the Arms Race: How the United States and the Soviet Union Develop New Military Technologies*. Ithaca, NY: Cornell University Press, 1988.

Fallows, James. *National Defense*. New York: Random House, 1981.

Fitzgerald, Frances. *Way out There in the Blue: Reagan, Star Wars, and the End of the Cold War*. New York: Simon & Schuster, 2000.

Geiger, Roger L. *Research and Relevant Knowledge: American Research Universities since World War II*. New York: Oxford University Press, 1993.

Gibson, James Norris. *The History of the U.S. Nuclear Arsenal*. Greenwich, CT: Brompton Books, 1989.

Gibson, James W. *The Perfect War: Technowar in Vietnam*. Boston and New York: Atlantic Monthly Press, 1986.

Gray, Chris Hables. *Postmodern War: The New Politics of Conflict*. New York: Guilford Press, 1997.

Hacker, Barton C. *Elements of Controversy: The Atomic Energy Commission and Radiation Safety in Nuclear Weapons Testing, 1947–1974*. Berkeley: University of California Press, 1994.

Hall, R. Cargill. *A History of the Military Polar Orbiting Meteorological Satellite Program*. Washington, DC: National Reconnaissance Office, 2001.

Hammond, Grant T. *The Mind at War: John Boyd and American Security*. Washington, DC: Smithsonian Institution Press, 2001.

Hartcup, Guy. *The Silent Revolution: The Development of Conventional Weapons, 1945–85*. London: Brassey's, 1993.

Herken, Gregg. *Counsels of War*. New York: Alfred A. Knopf, 1985.

Hewlett, Richard G., and Francis Duncan. *Nuclear Navy, 1946–1962*. Chicago: University of Chicago Press, 1974.

Hounshell, David A. *The Cold War, RAND, and the Generation of Knowledge, 1946–1962*. Santa Monica, CA: RAND Corporation, 1998.

Huisken, Ronald. *The Origin of the Strategic Cruise Missile*. New York: Praeger, 1981.

Kaplan, Fred. *The Wizards of Armageddon*. New York: Simon & Schuster, 1983.

Keaney, Thomas A., and Eliot A. Cohen. *Revolution in Warfare? Air Power in the Persian Gulf*. Annapolis, MD: Naval Institute Press, 1995.

Kelly, Martin Campbell, and William Aspray. *Computers: A History of the Information Machine*. New York: Basic Books, 1996.

Klare, Michael T. *War without End: American Planning for the Next Vietnams*. New York: Vintage, 1972.

Kleinman, Daniel Lee. *Politics on the Endless Frontier: Postwar Research Policy in the United States*. Durham, NC: Duke University Press, 1995.

Komer, Robert W. *Bureaucracy at War: U.S. Performance in the Vietnam Conflict*. Boulder, CO: Westview Press, 1986.

Lakoff, Sanford, and Herbert F. York. *A Shield in Space? Technology, Politics, and the Strategic Defense Initiative*. Berkeley: University of California Press, 1989.

Leslie, Stewart W. *The Cold War and American Science: The Military-Industrial-Academic Complex at MIT and Stanford*. New York: Columbia University Press, 1993.

Light, Jennifer S. *From Warfare to Welfare: Defense Intellectuals and Urban Problems in Cold War America*. Baltimore: Johns Hopkins University Press, 2003.

Lowen, Rebecca S. *Creating the Cold War University: The Transformation of Stanford.* Berkeley: University of California Press, 1997.

Mackenzie, Donald. *Inventing Accuracy: A Historical Sociology of Nuclear Missile Guidance.* Cambridge, MA: MIT Press, 1990.

Maneles, Mark David. *The Development of the B-52 and Jet Propulsion: A Case Study in Organizational Innovation.* Maxwell Air Force Base, Montgomery, AL: Air University Press, 1998.

McCrea, Frances B., and Gerald E. Markle. *Minutes to Midnight: Nuclear Weapons Peace Protest in America.* Newbury Park, CA: Sage, 1989.

McDougall, Walter A. *The Heavens and the Earth: A Political History of the Space Age.* New York: Basic Books, 1985.

Melman, Seymour. *Pentagon Capitalism: The Political Economy of War.* New York: McGraw-Hill, 1970.

Miller, Jerry. *Nuclear Weapons and Aircraft Carriers: How the Bomb Saved Naval Aviation.* Washington, DC: Smithsonian Institution Press, 2001.

Morris, Charles R. *Iron Destinies, Lost Opportunities: The Arms Race between the U.S.A. and the U.S.S.R., 1945–1987.* New York: Harper & Row, 1988.

Needell, Allan A. *Science, Cold War and the American State: Lloyd V. Berkner and the Balance of Professional Ideas.* Amsterdam: Harwood Academic, with National Air and Space Museum, Smithsonian Institution, 2000.

Neel, Spurgeon. *Medical Support of the U.S. Army in Vietnam, 1965–1970.* Vietnam Studies. Washington, DC: Department of the Army, 1991.

Neufeld, Jacob. *The Development of Ballistic Missiles in the United States Air Force, 1945–1960.* Washington, DC: Office of Air Force History, 1989.

Norberg, Arthur L., and Judy E. O'Neill. With contributions by Kerry J. Freedman. *Transforming Computer Technology: Information Processing for the Pentagon, 1962–1986.* Baltimore: Johns Hopkins University Press, 1996.

Ordway, Frederick I., III, and Mitchell Sharpe. *The Rocket Team.* New York: Thomas Y. Crowell, 1979.

Peebles, Curtis. *Guardians: Strategic Reconnaissance Satellites.* Novato, CA: Presidio Press, 1987.

Powaski, Ronald A. *March to Armageddon: The United States and the Nuclear Arms Race, 1939 to the Present.* New York: Oxford University Press, 1987.

Prokosch, Eric. *The Technology of Killing: A Military and Political History of Cluster Weapons.* London: Zed Books, 1995.

Redmond, Kent C., and Thomas M. Smith. *From Whirlwind to MITRE: The R&D Story of the SAGE Air Defense Computer.* Cambridge, MA: MIT Press, 2000.

Rich, Ben R., and Leo Janos. *Skunk Works: A Personal Memoir of My Years at Lockheed.* Boston: Little, Brown, 1994.

Richelson, Jeffrey T. *America's Secret Eyes in Space: The U.S. Keyhole Spy Satellite Program.* New York: Harper & Row, 1990.

———. *America's Space Sentinels: DSP Satellites and National Security.* Lawrence: University Press of Kansas, 1999.

Rip, Michael Russell, and James M. Hasik. *The Precision Revolution: GPS and the Future of Aerial Warfare.* Annapolis, MD: Naval Institute Press, 2002.

Rochlin, Gene I. *Trapped in the Net: The Unanticipated Consequences of Computers.* Princeton, NJ: Princeton University Press, 2000.

Roland, Alex. *The Military-Industrial Complex.* Washington, DC: American Historical Association, 2001.

Sapolsky, Harvey M. *The Polaris System Development: Bureaucratic and Programmatic Success in Government.* Cambridge, MA: MIT Press, 1972.

———. *Science and the Navy: The History of the Office of Naval Research.* Princeton, NJ: Princeton University Press, 1990.

Simpson, Christopher. *Universities and Empire: Money and Politics in the Social Sciences during the Cold War.* New York: New Press, 1998.

Simpson, John. *The Independent Nuclear State: The United States, Britain and the Military Atom.* New York: St. Martin's Press, 1983.

Spick, Mike. *Jet Fighter Performance: Korea to Vietnam.* London: Ian Allan, 1986.

Stine, G. Harry. *ICBM: The Making of the Weapon That Changed the World.* New York, Orion Books, 1991.

Thompson, Warren. *Korea: The Air War.* London: Octopus Books, 1992.

Thompson, Wayne, and Bernard C. Nalty. *Within Limits: The U.S. Air Force and the Korean War.* Washington, DC: Air Force History and Museums Program, 1996.

Werrell, Kenneth P. *The Evolution of the Cruise Missile.* Maxwell Air Force Base, Montgomery, AL: Air University Press, 1985.

Westrum, Ron. *Sidewinder: Creative Missile Development at China Lake.* Annapolis, MD: Naval Institute Press, 1999.

Westwick, Peter J. *The National Labs: Science in an American System, 1947–1974.* Cambridge, MA: Harvard University Press, 2003.

Wittner, Lawrence S. *The Struggle against the Bomb.* 2 vols. Stanford, CA: Stanford University Press, 1993–1997.

Wragg, David. *Helicopters at War: A Pictorial History.* New York: St. Martin's Press, 1983.

Wulforst, Harry. *Breakthrough to the Computer Age.* New York: Scribner's, 1982.

Yanarella, Ernest J. *The Missile Defense Controversy: Strategy, Technology, and Politics, 1955–1972.* Lexington: University of Kentucky Press, 1977.

York, Herbert. *The Advisors: Oppenheimer, Teller, and the Superbomb.* Stanford, CA: Stanford University Press, 1989. Reprint of 1976 edition.

Ziegler, Charles A. *Spying without Spies: Origins of America's Secret Nuclear Surveillance System.* Westport, CT: Praeger, 1995.

Index

About the Authors

BARTON C. HACKER is curator of Military History at the Smithsonian's National Museum of American History. He is a recipient of the Leonardo da Vinci Medal of the Society for the History of Technology and of several writing prizes. He has curated major exhibits on Submarines in the Cold War and West Point in the Making of America. His publications include books on the history of Project Gemini, radiation safety in nuclear weapons testing, and world military institutions, as well as many articles and book reviews on a wide range of topics in the history of military technology.

MARGARET VINING is Curatorial Specialist in Military History at the Smithsonian's National Museum of American History. She is also Secretary-General of the U.S. Commission on Military History. She has curated or co-curated major exhibits on the G.I. in World War II, Submarines in the Cold War, and West Point in the Making of America, as well as smaller exhibits on the G.I. Bill, the buffalo soldiers in the West, and navy women in World War I. Her publications include a book on West Point plus articles and reviews on military uniforms and women's military history.